PCR GURU

PCR GURU
An Ultimate Benchtop Reference for Molecular Biologists

AYAZ NAJAFOV
Harvard Medical School, Department of Cell Biology,
Boston, MA, United States

GERTA HOXHAJ
Harvard T.H. Chan School of Public Health,
Department of Genetics and Complex Diseases,
Boston, MA, United States

Amsterdam • Boston • Heidelberg • London
New York • Oxford • Paris • San Diego
San Francisco • Singapore • Sydney • Tokyo

Academic Press is an imprint of Elsevier

Academic Press is an imprint of Elsevier
125 London Wall, London EC2Y 5AS, United Kingdom
525 B Street, Suite 1800, San Diego, CA 92101-4495, United States
50 Hampshire Street, 5th Floor, Cambridge, MA 02139, United States
The Boulevard, Langford Lane, Kidlington, Oxford OX5 1GB, United Kingdom

Notices
Knowledge and best practice in this field are constantly changing. As new research and experience
broaden our understanding, changes in research methods, professional practices, or medical treatment
may become necessary.

Practitioners and researchers must always rely on their own experience and knowledge in evaluating and
using any information, methods, compounds, or experiments described herein. In using such information or methods they should be mindful of their own safety and the safety of others, including parties
for whom they have a professional responsibility.

To the fullest extent of the law, neither the Publisher nor the authors, contributors, or editors, assume
any liability for any injury and/or damage to persons or property as a matter of products liability,
negligence or otherwise, or from any use or operation of any methods, products, instructions, or ideas
contained in the material herein.

Library of Congress Cataloging-in-Publication Data
A catalog record for this book is available from the Library of Congress

British Library Cataloguing-in-Publication Data
A catalogue record for this book is available from the British Library

ISBN: 978-0-12-804231-1

For information on all Academic Press publications
visit our website at https://www.elsevier.com/

  Working together
to grow libraries in
developing countries

www.elsevier.com • www.bookaid.org

Publisher: Sara Tenney
Acquisitions Editor: Sara Tenney
Editorial Project Manager: Fenton Coulthurst
Production Project Manager: Chris Wortley
Designer: Christian Bilbow

Typeset by Thomson Digital

I dedicate this book to my father, Azad Najafov, who introduced me to the world of molecular biology and to my friend and mentor, İbrahim Barış, who taught me how to do my first PCR.

Ayaz Najafov

I dedicate this book to my family to whom I am eternally grateful for their constant support and care for me and my academic career.

Gerta Hoxhaj

CONTENTS

BIOGRAPHY

AYAZ NAJAFOV

Ayaz was born in Baku, Azerbaijan and as a high school student, participated in local and international biology Olympiads, where he had won several gold, silver, and bronze medals. He obtained his BSc and MSc degrees in Molecular Biology and Genetics from Bosphorus University, Istanbul, Turkey, and got a 2-year training in Biochemistry as an intern at Baylor College of Medicine, Houston, TX, United States. Ayaz pursued his PhD in Biochemistry at MRC Protein Phosphorylation Unit at Dundee University, Scotland, United Kingdom in Prof. Dario Alessi's lab. Ayaz is currently a postdoctoral fellow in Prof. Junying Yuan's lab at Harvard Medical School, Cell Biology Department.

GERTA HOXHAJ

Gerta was born in Fier, Albania. She obtained her BSc degree in Molecular Biology and Genetics, as well as a second major in Chemistry from Bosphorus University, Istanbul, Turkey. Gerta pursued her PhD in Biochemistry at MRC Protein Phosphorylation Unit at Dundee University, Scotland, United Kingdom in Prof. Carol MacKintosh's lab. Gerta is currently a postdoctoral fellow in Prof. Brendan Mannings's lab at Harvard T.H. Chan School of Public Health, Department of Genetics and Complex Diseases.

PREFACE

This book was written with an aim of providing researchers practicing molecular biology, whether at their beginner or expert stage, with a handy tool and reference for setting up, performing, optimizing, and troubleshooting polymerase chain reaction (PCR). We believe that this book will provide tremendous assistance for solving problems associated with challenging cases of PCR.

This book does not provide the reader with vast details of the PCR theory and the multiple applications of the technique. As this book is a *laboratory* guide, the emphasis here was made on the technical aspects of employing PCR as a tool in molecular biology laboratories and the issues were described in "bench terms," which makes the book more comprehendible to the end users of the technique.

General optimization and troubleshooting strategies, as well as a detailed guide for addressing specific PCR problems, tips and tricks developed and learned through years of experience with PCR and its various applications, several special cases for PCR, and appendices with protocols and other useful information related to PCR, make this book a practical reference all molecular biology laboratories can benefit from.

ACKNOWLEDGMENTS

We would like to thank Ali Kavak, Aydin Mammadov, and Cafer Ozdemir for their comments and suggestions during the preparation of this book.

NOTE TO THE READER

All agarose gel images published in this book were inverted (negative) to save ink/toner used to publish this book.

As DNA polymerase is the only enzyme in the standard PCR setup, and as there are many other DNA polymerases (besides Taq) used in modern molecular biology labs, we will refer to DNA polymerase as "the enzyme."

We will use the term "target" for the specific sequence that is intended to be amplified using a set of primers.

CHAPTER 1

Introduction

Abstract

Origins of polymerase chain reaction (PCR) and its history are briefly introduced in this chapter. The components of the PCR are introduced and the effect of the component concentrations on the fidelity, specificity, and efficiency of PCR are explained. Various thermostable DNA polymerases that are currently being used in the PCR setups around the world are listed. The power of the cyclic and the exponential nature of PCR, the effect of the reaction reaching the plateau, and how this affects the specificity and yield of the reaction are discussed. Finally the rise in the popularity of this method since it was first introduced by Dr. Kary Mullis in 1983 is illustrated.

For knowledge itself is power. (Francis Bacon)

Every great advance in science has issued from a new audacity of imagination. (John Dewey)

Real knowledge is to know the extent of one's ignorance. (Confucius)

1.1 A BIT OF HISTORY

Polymerase chain reaction (PCR) was invented by Dr. Kary Mullis in 1983. At that time, he was working at Cetus Corporation, one of the first biotechnology companies. For his invention, Dr. Mullis received a $10,000 bonus from Cetus. In 1992, Dr. Mullis sold the patent for PCR and Taq polymerase to Hoffmann La Roche for $300 million. In 1993, he received a Nobel Prize in Chemistry "for his invention of the polymerase chain reaction method" [1]. To read more about the history of PCR see Ref. [2].

1.2 WHAT IS PCR?

It is a rapid, relatively inexpensive, simple, sensitive, versatile, and robust in vitro method for specifically amplifying desired sequences of DNA or RNA molecules (Fig. 1.1). A large amount of products (up to milligrams of double-stranded DNA fragments) are generated at the end of each PCR, and a huge variety of downstream experiments and applications have been

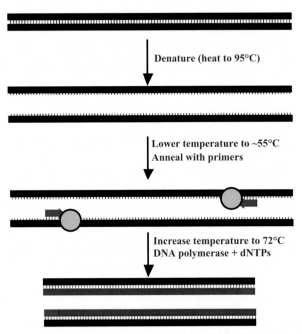

Figure 1.1 *Steps during the polymerase chain reaction (PCR).* DNA polymerase *(green)*, primers *(purple)*, and dNTPs *(pink)* are the main components of the reaction.

invented due to this technique's ability to generate double-stranded DNA at a large scale (Fig. 1.2). The applications of PCR include classification of organisms, genotyping, molecular archaeology, mutagenesis, mutation detection, sequencing, cancer research, detection of pathogens, DNA fingerprinting, drug discovery, genetic matching, genetic engineering, and prenatal diagnosis. The high applicability of PCR comes from its specificity and ability to detect even a single DNA molecule and produce millions of copies of the desired target sequence. As it is very hard, if not impossible, to manipulate DNA molecules at low quantities, by amplifying the DNA quantities to the levels that are very easy to handle, PCR revolutionized how people think about DNA experiments.

1.3 PCR COMPONENTS

There are six basic PCR components that are similar among all types of PCRs:

1. *Template DNA/RNA*: Contains the target sequence to be amplified. (e.g., genomic DNA, plasmid, amplicon, etc.).

2. *A pair of primers*: Oligonucleotides that define the sequence to be amplified. As the primer concentration increases, the reaction specificity decreases, while efficiency of the reaction increases.

> [primers] ↑ → Specificity ↓ and Efficiency ↑

3. *Deoxynucleotidetriphosphates (dNTPs)*: Are DNA-building blocks. As dNTP concentration increases, the reaction specificity and fidelity decreases, while efficiency of the reaction increases.

> [dNTP] ↑ → Specificity ↓ and Fidelity ↓ and Efficiency ↑

4. *Thermostable DNA polymerase*: Is the enzyme that catalyzes the reaction. Originally, the DNA polymerase employed in PCR was isolated from *Escherichia coli*. However, as PCR involves multiple cycles of temperatures that denature and inactivate the *E. coli* DNA polymerase, researchers had to manually add a new aliquot of the enzyme at the beginning of each cycle. Fortunately, soon after invention of PCR, a thermostable DNA polymerase, which is nowadays famously known as "Taq," was isolated from *Thermus aquaticus*, a bacterium that lives in hot springs. Later, thermostable DNA polymerases were isolated from many other bacterial strains (e.g., Pfu, Tma, and KOD). The optimum temperature for Taq activity is between 75 and 80°C [3] and half-life of this enzyme at 95°C is about 40 min. Taq does not have a proofreading activity (3′–5′ exonuclease), and thus the frequency of mutations incorporated into PCR products is about 1 in 200,000 per nucleotide per duplication of the DNA [4–6]. For applications where fidelity of the DNA polymerase is critical, polymerases such as Pfu and Pwo are used. These DNA polymerases have a proofreading activity (3′–5′ exonuclease) and a fidelity that is about 10 times higher than the fidelity of Taq DNA polymerase [5,6]. For detailed comparison of the thermostable DNA polymerases, see Appendix I.

    The concentration of enzymes is defined in units. Each enzyme-supplying company provides the definition of a unit for the enzyme it supplies. For example, New England Biolabs (NEB) defines 1 U of Taq as: the amount of enzyme that incorporates 10 nmols of dNTPs into acid–insoluble material in 30 min at 75°C. Usually, these definitions are close to each other, if one is switching to another company's enzyme, it is recommended to compare the definitions of units and the concentrations of the supplied enzymes, to have the optimized PCRs still running with the new enzyme too.

5. *Mg⁺⁺ ions*: Cofactor of the enzyme. As Mg^{++} concentration increases, the reaction specificity decreases, while efficiency of the reaction increases.

$[Mg^{++}]$ ↑ → Specificity ↓ and Efficiency ↑

6. *Buffer solution*: Maintains pH and ionic strength of the reaction solution suitable for the activity of the enzyme.

Specificity and efficiency of PCR depend on correctly balanced, optimum combination of the concentrations of these reagents. A standard PCR setup (see Chapter 2) is used for most purposes and the only factors that need to be optimized for each new pair of primers are the annealing temperature and Mg^{++} ion concentration. The rest of the parameters are very similar among most PCRs. Special cases, however, should be dealt with care and standard PCR protocols are vastly modified (see Chapter 6).

1.4 EXPONENTIAL NATURE OF PCR

The exponential accumulation of the products is one of the most important features of PCR. If one begins with 1 molecule as a template, at the end of 30 cycles, theoretically, around 2^{30} (1,073,741,824) molecules should be present in the PCR tube. This amplification scale, in fact, is one of the major factors that made this technique so popular and crucial in the modern molecular biology world.

However, this exponential accumulation of the products does not continue forever. After an exponential phase that lasts about 30 cycles (in most cases), a plateau phase is observed and the product amplification here is about linear (Fig. 1.2).

The reasons for this plateau stage are thought to be:
- degradation of reactants (dNTPs and Taq),
- depletion of reactants (dNTPs and primers),
- reaction inhibition by the end product (pyrophosphate),
- reaction inhibition by competition of reactants for nonspecific products, and
- reaction inhibition due to easier reannealing of concentrated products and decrease in primer binding.

In addition to poor accumulation of products, this plateau phase is known to generate nonspecific products. Thus, the number of cycles is usually set to no more than 35.

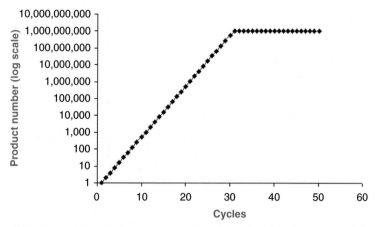

Figure 1.2 *A linear plateau phase appears about 30 cycles after the exponential phase.*

1.5 POPULARITY OF PCR

After the first PCR articles in 1985, the popularity of this technique sky-rocketed. Fig. 1.3 depicts the number of articles retrieved per year when the term "polymerase chain reaction" is searched in PubMed. If these article numbers are added (Fig. 1.4), there is an almost purely exponential accumulation of number of articles that can found through PubMed.

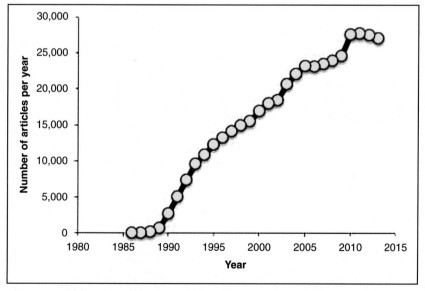

Figure 1.3 *The number of articles found per year, when the phrase "polymerase chain reaction" is searched in PubMed (http://www.ncbi.nlm.nih.gov/pubmed/).*

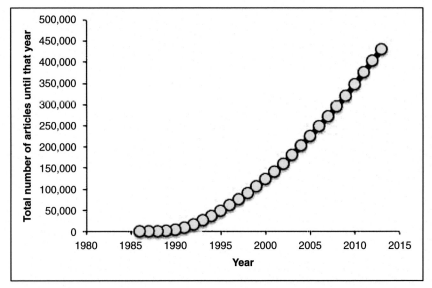

Figure 1.4 *The number of accumulated articles over the years, when the phrase "polymerase chain reaction" is searched in PubMed (http://www.ncbi.nlm.nih.gov/pubmed/).*

The most obvious reasons for this increase in popularity are the low cost and effectiveness of PCR. There are many techniques in the molecular biology world and many of them are expensive, and this factor limits the number of laboratories that can pursue these fancy techniques. However, a PCR machine and several inexpensive reagents are available to most laboratories in the world, which is why almost any molecular biology laboratory nowadays, has at least one thermocycler. It would have been totally unbelievable for researchers before mid-1980s, but PCR today is a molecular biology laboratory's essential core technique. Many labs are using PCR as their primary method of generating data.

Finally, PCR was one of the most important cornerstones of the human genome project.

CHAPTER 2

Procedure

Abstract

This chapter focuses on primer design and basics of polymerase chain reaction (PCR). Critical parameters to consider while designing primers are discussed. An example of primer design for a molecular cloning project using restriction enzymes is illustrated. Various modern online software resources for primer design and DNA sequence analysis are listed with illustrated outputs. A typical PCR setup is explained in detail and the recommended starting concentrations of the components, as well as usual ranges for optimization are tabulated. Default positive and negative controls for rigorous PCR setup are discussed. The temperature cycling parameters—denaturation, annealing, and extension—are explained in detail. How these parameters should be altered depending on the target DNA, and how these parameters affect specificity and yield of PCR are explained.

There's a difference between knowing the path, and walking the path.

(Morpheus)

We will not know unless we begin.

(Peter Nivio Zarlenga)

Knowing is not enough; we must apply. Willing is not enough; we must do.

(Johann von Goethe)

2.1 PRIMER DESIGN

It is always best to use published primer sequences because these primers have been previously tested by other researchers. However, if any previously published pair of primers is not present for a particular application, this section will provide the essential information needed required for primer designing.

Primer design is *the most* important variable in polymerase chain reaction (PCR) (Fig. 2.1). The success or failure of PCR reactions mostly depends on how good the primers are. Primer design quality determines the specificity and the efficiency of the PCR. Nevertheless, for most target sequences, it is fairly easy to design primers for simple amplification purposes.

The smartest way to design primers is by using software that takes into account various parameters of primer design, such as GC richness, annealing temperature of the primers (T_m), etc. There are many software packages

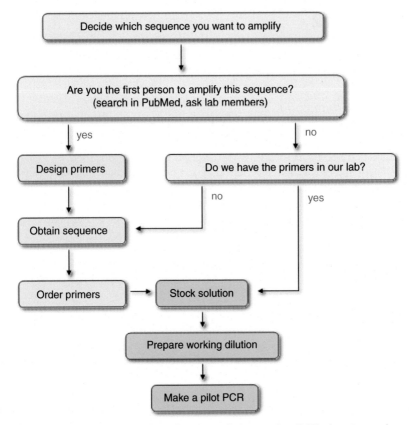

Figure 2.1 *A simple flowchart for polymerase chain reaction (PCR) planning and setup.*

that can be purchased for this purpose, but there are several free web-based tools (Table 2.1), which serve the same purpose just as well. These user-friendly online algorithms provide several choices of primer pair sequences for the target sequence one chooses as input. A range of parameters, such as GC content and T_m, can also be defined during the design process. T_m, molecular weight, %GC, and various other properties can be calculated for any given oligo sequence by using the tools listed in Table 2.1.

However, even though these primer design software do a pretty good job, their output has to be checked. The tools output several primer pairs, which are the best possible choices (as far as the algorithm can calculate), for the target sequence. In most cases one of these suggested primer pairs will work perfectly for a PCR. However, one needs to verify certain properties of the suggested sequences, before ordering them.

The main parameters [7] for verification are listed.

1. A primer should match only one sequence in the template.
2. Primer length should be 18–30 nucleotides long.
3. GC content should be 40–60%.
4. Primer pairs should have similar T_m values (±5°C).
5. The T_m values should be between 50 and 80°C (preferably).
6. Secondary structures should be absent.
7. Ts at the very 3′-ends should be absent.
8. 3′-ends should end with a G or C, or CG or GC.
9. 3′-ends should *not* end with three or more Cs or Gs.
10. Primers should not basepair with each other (especially in the 3′-end).

Some researchers manually select two short (18–30 nt) sequences that flank the target sequence and then check whether these potential primer sequences match the criteria. Oligo calculators listed in Table 2.1 can be used for this purpose.

Let's say the project is to clone Connexin32 (Cx32) protein into a plasmid vector. Here is what one would do:

Step 1: Get the coding sequence

First, find the protein's mRNA entry in Entrez/Nucleotide (http://www.ncbi.nlm.nih.gov/entrez/query.fcgi?db=Nucleotide)

Search for "Connexin 32" in the database.

Usually, the first listed entry is what one is looking for. Be careful not to get confused with the species. Type "connexin 32 *Homo sapiens*" in the query box if it is a human protein that is to be searched.

Click on the entry to get the important information about the protein: sequence, accession, and GI number. Save this information, which might be required further.

Table 2.1 Free online tools for primer design and oligo property analysis

Tool	Hyperlink
OligoPerfect Designer	http://www.invitrogen.com/content.cfm?pageid=9716
Primer3	http://bioinfo.ut.ee/primer3/
Web Primer	http://seq.yeastgenome.org/cgi-bin/web-primer
Simple Oligo Calculator	http://www.sciencelauncher.com/oligocalc.html
Oligo Properties Calculator	http://www.basic.northwestern.edu/biotools/oligocalc.html

OUTPUT:

Homo sapiens gap junction protein, beta 132 kDa (connexin 32, Charcot–Marie–Tooth neuropathy, X-linked) (GJB1).

ACCESSION:	NM_000166
GI:	31542846
mRNA:	1638 bp

Start codon

```
   1 gcggtgatga attgggacgc aggcgcggag cccagggacc actccccctg cacagacatg
  61 agaccatagg ggacctgtct gggtggcctc agggataggc gctccccaag gtgtgaatga
 121 ggcaggatga actgctcagg tttgtacacc ttgctcagtg gcgtgaaccg gcattctact
 181 gccattggcc gagtatggct ctcggtcatc ttcatcttca gaatcatggt gctggtggtg
 241 gctgcagaga gtgtgtgggg tgatgagaaa tcttccttca tctgcaacac actccagcct
 301 ggctgcaaca gcgtttgcta tgaccaattc ttccccatct cccatgtgcg gctgtggtcc
 361 ctgcagctca tcctagtttc caccccagct ctcctcgtgg ccatgcacgt ggctcaccag
 421 caacacatag agaagaaaat gctacggctt gagggccatg gggaccccct acacctggag
 481 gaggtgaaga ggcacaaggt ccacatctca gggacactgt ggtggaccta tgtcatcagc
 541 gtggtgttcc ggctgtttgtt tgaggccgtc ttcatgtatg tcttttatct gctctaccct
 601 ggctatgcca tggtgcggct ggtcaagtgc gacgtctacc cctgccccaa cacagtggac
 661 tgcttcgtgt cccgccccac cgagaaaacc gtcttcatgc tctttcatgct agctgcctct
 721 ggcatctgca tcatcctcaa tgtggccgag gtggtgtacc tcatcatccg ggcctgtgcc
 781 cgccgagccc agcgccgctc caatccacct tcccgcaagg gctcgggctt cggccaccgc
 841 ctctcacctg aatacaagca gaatgagatc aacaagctgc tgagtgagca ggatggctcc
 901 ctgaaagaca tactgcgccg cagccctggc accgggggctg ggctggctga aaagagcgac
 961 cgctgctcgg cctgctgctg ccacatacca ggcaacctcc catcccaccc ccgaccctgc
1021 cctgggcgag ccctccttc ttcctgccgg gtgcacggc ctctgcctgc tggggattac
1081 tcgatcaaaa ccttccttcc ctggctactt cccttcctcc cggggccttc cttttgagga
1141 gctggagggg tggggagcta gaggccacct atgccagtgc tcaaggttac tgggagtgtg
1201 ggctgccctt gttgcctgca cgcttccctc ttccctcctc ctctctctgg gaccactggg
1261 tacaagagat gggatgctcc gacagcgtct ccaattatga aactaatctt aaccctgtgc
1321 tgtcagatac cctgtttctg gagtcacatc agtgaggagg gatgtgggta agaggagcag
1381 agggcagggg tgctgtggac atgtggggtg agaaggaggg gtggccagca ctagtaaagg
1441 aggaatagtg cttgctggcg acaaggaaaa ggaggaggtg tctggggtga gggagttagg
1501 gagagagaag caggcagata agttggagca ggggttggtc aaggccacct ctgcctctag
1561 tccccaaggc ctctctctctc ctgaaatgtt acacattaaa caggatttta cagccaaaaa
1621 aaaaaaaaaa aaaaaaaa
```

Stop codon

Underlines nucleotides = primer matches

Step 2: Get the amino acid sequence

The Entrez/Protein database contains the protein's amino acid sequence. (http://www.ncbi.nlm.nih.gov/entrez/query.fcgi?db=Protein)

OUTPUT:

Protein ID:	NP_000157
Length:	283 aa

```
  1   mnwtglytll sgvnrhstai grvwlsvifi frimvlvvaa esvwg dekss ficntlqpgc
 61   nsvcydqffp ishvrlwslq lilvstpall vamhvahqqh iekkmlrleg hgdplhleev
121   krhkvhisgt lwwtyvisvv frllfeavfm yvfyllypgy amvrlvkcdv ypcpntvdcf
181   vsrptektvf tvfmlaasgi ciilnvaevv yliiracarr aqrrsnppsr kgsgfghrls
241   peykqneink llseqdgslk dilrrspgtg aglaeksdrc sac
```

Step 3: Get the ORF

Now determine the open reading frame (ORF) of the protein in the mRNA sequence. In other words, find the start and the stop codons (the output from the Entrez/Nucleotide will not provide the ORF information).

To get the ORF, align mRNA and protein sequences at: http://blast.ncbi.nlm.nih.gov/

by using program blastx with the following parameters:

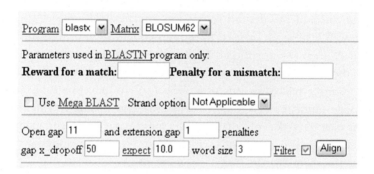

OUTPUT:

Scheme of the alignment (blue box = protein sequence match on the mRNA):

```
Alignment:
Query: 127  MNWTGLYTLLSGVNRHSTAIGRVWLSVIFIFRIMVLVVAAESVWGDEKSSFICNTLQPGC  306
            MNWTGLYTLLSGVNRHSTAIGRVWLSVIFIFRIMVLVVAAESVWGDEKSSFICNTLQPGC
Sbjct: 1    MNWTGLYTLLSGVNRHSTAIGRVWLSVIFIFRIMVLVVAAESVWGDEKSSFICNTLQPGC  60

Query: 307  NSVCYDQFFPISHVRLWSLQLILVSTPALLVAMHVAHQQHIEKKMLRLEGHGDPLHLEEV  486
            NSVCYDQFFPISHVRLWSLQLILVSTPALLVAMHVAHQQHIEKKMLRLEGHGDPLHLEEV
Sbjct: 61   NSVCYDQFFPISHVRLWSLQLILVSTPALLVAMHVAHQQHIEKKMLRLEGHGDPLHLEEV  120

Query: 487  KRHKVHISGTLWWTYVISVVFRLLFEAVFMYVFYLLYPGYAMVRLVKCDVYPCPNTVDCF  666
            KRHKVHISGTLWWTYVISVVFRLLFEAVFMYVFYLLYPGYAMVRLVKCDVYPCPNTVDCF
Sbjct: 121  KRHKVHISGTLWWTYVISVVFRLLFEAVFMYVFYLLYPGYAMVRLVKCDVYPCPNTVDCF  180

Query: 667  VSRPTEKTVFTVFMLAASGICIILNVAEVVYLIIXXXXXXXXXXXSNPPSRKGSGFGHRLS  846
            VSRPTEKTVFTVFMLAASGICIILNVAEVVYLII           SNPPSRKGSGFGHRLS
Sbjct: 181  VSRPTEKTVFTVFMLAASGICIILNVAEVVYLIIRACARRAQRRSNPPSRKGSGFGHRLS  240

Query: 847  PEYKQNEINKLLSEQDGSLKDILRRSPGTGAGLAEKSDRCSAC  975
            PEYKQNEINKLLSEQDGSLKDILRRSPGTGAGLAEKSDRCSAC
Sbjct: 241  PEYKQNEINKLLSEQDGSLKDILRRSPGTGAGLAEKSDRCSAC  283
```

Alternatively, the insert cDNA can be obtained from various company sites where the constructs are being sold (e.g., origene.com or addgene.com).

Step 4: Design primers

Primers can be "designed" by picking any sequence flanking the ORF, but as PCR should be easy to optimize, it is better to *design* primers by taking certain factors have into account.

Primers are nowadays designed by using various online and offline software. We suggest OligoPerfect Designer primer design tool: http://www.invitrogen.com/content.cfm?pageid=9716

This tool can add the desired restriction enzyme (RE) sites into the primers. These RE sites can be used in the subsequent steps of the cloning.

Important Note 1: Do not forget to add three to four random nucleotides before the RE sites, to allow the REs to bind to the DNA efficiently. See NEB's resource related to this issue: http://www.neb.com/tools-and-resources/usage-guidelines/cleavage-close-to-the-end-of-dna-fragments)

Important Note 2: Make sure that the RE sites inserted into the primers (which is decided by the vector into which the insert will be cloned in) are absent in the insert, so that the insert is not cut in an unwanted manner.

NEBCutter (http://tools.neb.com/NEBcutter2/) can be used for this purpose, and the output will appear something like this:

OUTPUT:

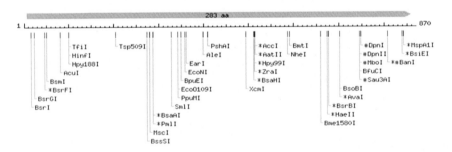

Cloning primer architecture
The final primer sequences are:

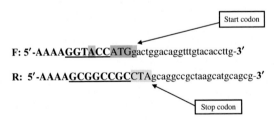

F: 5′-AAAAGGTACCATGgactggacaggtttgtacaccttg-3′

R: 5′-AAAAGCGGCCGCCTAgcaggccgctaagcatgcagcg-3′

KpnI and NotI sites (underlined) were included for directional cloning and AAAA spacer nucleotides were added to the primers' 5'-ends to allow REs to bind to the DNA efficiently. The Kozak sequence (which in this case overlaps with the KpnI site) is highlighted in blue.

OligoPerfect Designer tool will provide the designed primer's properties. If for any reason, one needs to recheck T_m and GC% of the primers, this tool can be used:

http://sciencelauncher.com/oligocalc.html

OUTPUT:

For the forward primer:

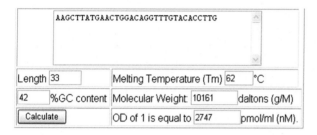

For the reverse primer:

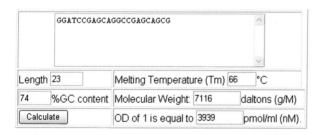

Important Note 3: If there is no N-terminal tag in the vector where the insert is to be cloned, a Kozak sequence (**[G/A]**NNATG**G**) should be added into the forward primer just before the start codon (ATG) to ensure proper translation initiation. Using a KpnI RE site in the forward primer just before the ATG automatically meets this requirement.

A final checklist for primer design (for cloning):
• Do the picked REs cut the insert in an unwanted place?
• Is the start codon put in frame with the ORF?
• Is the stop codon put in frame with the ORF?

- Is the "stuffer" (3–4 nucleotides at the oligo 5′-terminus for RE binding) put into the primer?
- Is the Kozak sequence put into the primer?
- Is the insert going to be in frame with the N- or C-terminal tags in the vector?
- Is there an unwanted stop codon that would prevent C-terminal tagging from vector?

2.2 STANDARD PCR SETUP

Optimal conditions for a particular PCR mainly vary when primer pairs and template DNAs are changed. However, even slight changes in any other component might perturb the successful setup and might lead to loss of quality or even total lack of amplification in a PCR setup that was previously giving perfect results. Usually, PCR setups are similar within the template categories (genomic DNA, plasmids, and cDNA) and differ only among various primer pairs. DNA polymerase is also an important factor. If a new kind of polymerase (e.g., Taq vs. Pfu) is used or the same Taq is purchased from another company, the previous PCR setup might or might not yield acceptable results and thus a new optimization may be necessary.

Nevertheless, the concentration ranges for many standard PCRs (see Chapter 6 for special cases) are almost universal (Table 2.2). For instance, the

Table 2.2 Concentrations to start with (and then optimize if necessary) and concentration ranges (within which most optimizations should be done) for the standard PCR components

Component	Recommended starting concentrations	Usual ranges for the final concentrations
ddH2O	—	—
10× Buffer	1×[a]	1×[a]
MgCl$_2$	1 mM	0.5–4.0 mM
Forward primer	0.5 μM	0.4–1.5 μM
Reverse primer	0.5 μM	0.4–1.5 μM
Template DNA	0.5 ng/μL (plasmid) 5 ng/μL (genomic)	0.1–1.0 ng/μL (plasmid) 1–10 ng/μL (genomic)
dNTPs	0.25 mM	0.2–0.5 mM
DNA polymerase	0.05 U/μL	0.025–0.05 U/μL

[a]Usually 1× is 50 mM KCl, 10 mM Tris–HCl, pH 8.3. However, various enzyme suppliers suggest various buffers. The best is to use the buffer suggested and provided by the enzyme supplier. dNTPs, Deoxynucleotidetriphosphates.

most common concentration for deoxynucleotidetriphosphates (dNTPs) is 0.2 mM. Buffer is supposed to be 1× and the DNA polymerase concentration varies according to the brand and type of polymerase and is always suggested by the manufacturer. Still, the usual range for most Taq DNA polymerases is somewhat within the range of 0.02–0.05 U/μL.

It is important to accurately calculate how many microliters should be mixed to achieve these concentrations in the final reaction volume. An example recipe for a 100-μL reaction mix (enough for four to five individual reactions, 20–25 μL each) is:

Sample 100-μl reaction mix:

dH$_2$O	→ 82.5 μL
10× Buffer	→ 10 μL
50 mM MgCl$_2$	→ 2 μL
12.5 mM dNTPs	→ 2 μL
50 μM Forward primer	→ 1 μL
50 μM Reverse primer	→ 1 μL
1 μg/μL Template	→ 0.5 μL
5 U/μL Taq	→ 1 μL

The final volume of the reaction depends on the downstream application. However, it is important to not overload the tubes to ensure uniform heating and cooling of the tube during the PCR cycles. For *0.2-mL* tubes volumes not larger than *50 μL*, and for *0.5-mL* tubes volumes not larger than *100 μL* should be used.

2.3 CONTROLS

Inclusion of various positive and negative controls is critical for the correct interpretation of the results obtained on an agarose gel. This is especially important if the PCR setup does not work and troubleshooting is required.

Controls are additional tubes containing everything but a component, which requires troubleshooting. Depending on the experiment, a control tube might be missing the primers, dNTPs, or enzyme. These controls are usually performed when something is going wrong with the PCR and helps in figuring out the problem.

Some of the common controls are:
1. Negative control
   a. That is, "no DNA control": Add the same volume of dH$_2$O instead of the template.

    b. Negative control is used to make sure that the amplification is specific and that there is no contamination. If there is a band in the negative control refer to Chapter 4.

2. Positive control:

    a. That is, a "control that works": Add a previously amplified amplicon or a template that is known to have the target sequence instead of the regular template that is used for the rest of the samples in a particular PCR setup.

    b. Positive control is used to make sure that there is nothing wrong with the PCR setup. In case a PCR yields no bands in all the samples except for the positive control, one can make a judgment about the template used in the rest of the samples. Positive control is a reference to which one can compare all the samples. One can think of the positive control as "this is how the bands should look when the target sequence is present."

    c. If for instance, in a previously optimized PCR, a new kind of template is used and none of samples (except for positive control) yielded a product, one can safely judge that there is a problem with the new template. If a positive control was not present in this PCR, this judgment could not have been made because there would be no way of finding out if PCR worked.

2.4 CYCLING PARAMETERS

The three steps of PCR (denaturation, annealing, and extension) usually change from setup to setup. However for the same type of PCR (e.g., amplification for subsequent mutation detection step or colony PCR), these steps almost always remain the same and only the annealing temperature changes.

2.4.1 Denaturation

Denaturation length is usually 0.5–2.0 mins and the temperature is usually 94–95°C. This length and time depends on the size of the template (genomic vs. cDNA vs. plasmids) and the GC richness of the template. The larger is the template and the higher is the %GC, the longer and/or the higher the denaturation step should be. The effect of the denaturation step is expressed in terms of length × temperature. So, when optimization is necessary, either the length of the denaturation or the temperature can be increased, but the temperature should not be higher than 96°C. Initially, a setting of *94°C/15 s* should be tried. Depending on the output of the reaction, this setting might need changing (see Chapter 3). Also, the denaturation step may depend on the DNA polymerase enzyme used, thus consult the manufacturer's instructions for the best temperature and time.

In some cases, it is better to set the denaturation step for 10 s at 95°C. This is because exposure of DNA to high temperatures for long periods may lead to depurination of single-stranded DNA during and subsequent strand scission. Also, prolonged exposure of the thermostable DNA polymerase to high temperatures results in a gradual loss of enzyme activity. Keeping the denaturation step short is especially important when amplifying long target sequences because the total PCR cycling time will be several hours.

Note: Before the first cycle starts, there is an initial *94°C/4–5 min* denaturation step that helps to make sure no potentially harmful enzymes (e.g., DNases) are active. Also, this step ensures complete denaturation of the template DNA. Some researchers find this unnecessary, so this step can be omitted if one would like to save 5 min.

2.4.2 Annealing

Annealing setting is almost always $(T_m - 4)/30$ *s*. Only the temperature changes among primer sets. The widely practiced standard is to set the annealing temperature in the first PCR with a new set of primers to a $T_m - 4$ value. In other words, if the T_m is 64°C, the first PCR should be performed at an annealing temperature = 60°C.

Note: T_m is the theoretical value, which is defined as the temperature at which an oligo will be half-denatured (given that it was a double-stranded DNA). It is a theoretical value (i.e., calculated) and the actual value depends on many factors, including ionic strength. Tools listed in Table 2.1 can be used to determine the T_m of any given DNA sequence.

2.4.3 Extension

Extension temperature can be said to be universal: *72°C*. However, as for the denaturation step, check the manual of the DNA polymerase supplier, as some polymerases perform optimally at 68°C. The length of the extension step depends on the length of the target sequence to be amplified (Table 2.3).

At the end of the whole PCR cycling program, there is usually a *72°C/5–10 min* step. This step allows polymerases to finish their job completely and it also reduces nonspecific background, in some cases.

Table 2.3 Extension times for target sequences of various lengths

Target Sequence (kb)	Extension time (min)
<1	0.5–1.0
1–5	3
6–9	6
10–20[a]	12

[a]See Chapter 6 for a descriptive discussion of "long PCR."

The whole PCR program usually looks something like this:
1. 94°C/5 min
2. 94°C/15 s ⎫
                       ⎬ 25–35 cycles
3. 55–57°C/30 s ⎭
4. 72°C/0.5–6.0 min
5. 72°C/5 min

The usual thermocycler setting is to repeat the steps 2–3 about 25–35 times. If the template is genomic DNA, it is better to set the number of cycles to 33–35 cycles. If the template is a plasmid, even 25 cycles would be sufficient in most cases (depends on the downstream application).

Note: A setting of more than 35 cycles will usually *not* yield more PCR product, but only more nonspecific bands. The plateau effect of PCR is the reason for this and it is discussed in Chapter 1.

CHAPTER 3

Good PCR Practices

Abstract

Thirty-three routine benchtop practices for successful polymerase chain reaction (PCR) are explained in detail in this chapter. Important aspects of a PCR routine, such as maintenance of cleanliness of the PCR environment and its decontamination; prevention of cross-contamination of the PCR samples; prevention of deterioration of PCR reagents and extension of their shelf life by appropriate aliquoting, storage, and handling; usage of correct tips and PCR racks; usage of PCR product size–appropriate agarose gel concentrations for checking the PCR results; tube labeling; record keeping; and accurate pipetting techniques are discussed. Methods for ensuring the template DNA used for PCR is of high quality are explained.

There is nothing so easy to learn as experience and nothing so hard to apply.

(Josh Billings)

What one has not experienced, one will never understand in print.

(Isadora Duncan)

Aside from knowing how to design a set of primers and how to set up a polymerase chain reaction (PCR) reaction, one should also know how to do PCR *properly*. In other words, what kind of rules and principles should be kept in mind to have a successful PCR result.

There is a big difference between a "PCR that does not work" because something went wrong or because a good PCR practice was not conducted, and a "PCR that did not work" because the reaction conditions just need to be optimized.

This chapter describes the basic guidelines that have to be followed to eliminate the "PCR that does not work" situations. However, if a satisfactory PCR result is not obtained after following all the rules described in this chapter, refer to the next chapter that is devoted to PCR troubleshooting.

Good Practice #1: Have the calculations done before putting on gloves for PCR.

It is a good practice to write down the recipes for each PCR that are to be done, instead of calculating how many microliters should be added. Even if one has performed a PCR thousand times before, it is always better to invest 30 s and write it down, rather than redoing the whole PCR again, just because something was miscalculated or forgotten to be included into the master mix. A little recipe written down will also help to keep track whether all the reagents were included or not.

Good Practice #2: Always prepare a master mix for n + 1 samples.

The pipettes are never 100% accurate. Even if, 10 times 10 is equal to 100 and 100 divided by 10 will give 100, this is not always strictly true for pipetting. There is always some minor pipetting error and some volume is always carried on the outer pipette tip walls and hence, 2 + 2 is never 4 in pipetting. Therefore, if PCR is to be performed for 10 samples, prepare the master mix for 11 samples. This will ensure that the issue of insufficient volume for the last sample will not arise and that all the samples will receive an equal volume of the master mix.

Good Practice #3: Check if all the necessary reagents are present before starting.

Sometimes one can run out of deoxynucleotidetriphosphates (dNTPs) or the enzyme or a primer solution and this is realized only after most of the reagents are added into the master mix. Keep in mind that some of the PCR reagents are not stable at room temperature and it is better to keep the setup time to a minimum. Just check if all the reagents are available and there is enough volume for the particular PCR that is about to set up. Making this a habit will prevent one from jeopardizing the quality of half-finished PCRs.

Good Practice #4: Use dedicated pipettes.

Every lab that performs PCR *should* have a set of pipettes, which is used exclusively for PCR. Use them only when dilutions for the PCR components (except template DNA dilutions) are to be prepared. Using pipettes that are used to check PCR results (e.g., when loading samples into the agarose gel) poses a great risk of contaminating current PCR samples or stock solutions with some of the PCR products. If not now, this might cause one or the other lab member to suffer greatly. If such a dedicated pipette set is not present, buy a new one or send a set for cleaning and calibration. Having only P2 (0.5–2.0 µL), P20 (2–20 µL), and P200 (20–200 µL) should be sufficient.

Good Practice #5: Use calibrated pipettes.

Using calibrated pipettes is just as important as using dedicated pipettes. Never use an uncalibrated pipette set. The results will just not be reproducible. In other words, one is not really doing science if one does not know whether a volume was really 5 µL or only 3 µL of a reactant that was put in a tube. Send pipettes for calibration and clean them regularly!

Good Practice #6: Work in a dedicated workspace with a dedicated lab coat.

It is best to have a hood dedicated to PCR, so called "PCR station." But if a separate hood is not available, try to use the same bench or the same side/corner of the bench for PCR and do not do any other experiments in that particular workspace. If possible, spray the area with 70% ethanol and/or UV irradiate the hood/bench for 10 min before use. Especially, keep the post-PCR samples (samples that are taken out of the thermocycler) away from the PCR station.

A clean lab coat that is used only during PCR is a good habit if PCR applications involve a low amount of template DNA and/or are susceptible to contamination with human genomic DNA (i.e., when working with primers for human genomic sequences).

Good Practice #7: Do not immerse pipette tips too deep into solutions.

The outer walls of the tip will take on some volume of the reactant and the actual volume put into the tube will be altered. This is particularly dangerous when working with viscous solutions, such as the enzyme solutions that have a high-glycerol concentration (50%) in them. One might think that an extra 20% volume of the Taq solution would not hurt the PCR setup, but this turns out to be *not* the case in practice. The extra glycerol that is added to the reaction tends to inhibit the reaction and thus the excess of enzyme solution is not a benefit, but a drawback here. For this reason, dip the pipette tip just deep enough to be able to take the liquid, but do not take air instead of the liquid!

Good Practice #8: Always thaw–vortex–spin the $MgCl_2$ solution.

Once the solutions are thawed, a concentration gradient forms inside the tube. This, of course, will lead to an inconsistency between experiments and may lead to a total lack of positive results due to the introduction of too low/high amounts of Mg^{++}. This is true for all of the chemicals thawed, but is more pronounced for $MgCl_2$.

Spinning the solution down to the bottom of the tube is critical for prevention of aerosols and potential contamination.

Good Practice #9: Always put on new gloves before starting a PCR setup.

Gloves that are just used for some other tasks in the lab will most likely have some sort of chemical or perhaps even DNA that can contaminate the PCR and lead to an inhibition or amplification of an incorrect DNA template. In case of human mutation research, one needs to be extra careful not to contaminate patient samples with PCR products and/or genomic DNA of other patients. If a door is opened with a glove, put on a new one before PCR-dedicated pipettes, PCR tubes, and reagents are touched.

Good Practice #10: Wipe the benchtop with EtOH.

Even if a dedicated space for PCR is present, it is a good idea to wipe the benchtop surface with 70% ethanol. Ethanol precipitates DNA and thus decreases the chances of contaminating a new PCR with some old DNA. Also, ethanol will get rid of microorganisms and other dirt.

Good Practice #11: Make sure the racks are clean.

Try to keep the racks for setting up PCR and for analyzing PCR separately. Dedicate at least one for the pre-PCR phase and use it exclusively for this purpose. Many different people use racks in the labs and various chemicals spill on those racks. Keeping the racks clean will prevent a PCR result from being negative just because the rack was dirty.

Good Practice #12: Make sure the tubes are clean.

Also, always use clean tubes. Most people autoclave PCR tubes and this is a good idea. But sometimes the suppliers of the tubes guarantee that the tubes are sterile upon opening of the package. In this case, it is not necessary to autoclave the tubes, but it would be just a good preliminary measure to have contamination-free PCR tubes. If tubes are autoclaved, make sure that the autoclave is itself clean. If somebody used it before and dirty things were autoclaved, it is a good idea to run the autoclave once without putting tubes in it. Theoretically, nothing that comes out of the autoclave should be dangerous to PCR, but this is just a preliminary measure that should be kept in mind.

Good Practice #13: Store dNTP in small aliquots and thaw them at the end.

dNTPs in an aqueous solution are not stable upon multiple thawing and freezing cycles, they tend to degrade. Even though they will eventually be heated and cooled about 30 times during PCR, it is best to store them at $-20°C$ in small aliquots, so that the freeze–thaw cycle for each aliquot is no more than 5 times. Calculate how much dNTP is used up in an average PCR setup and aliquot the dNTP solution according to that volume.

Also, when preparing the master mix, thaw dNTPs at the end (try to take them out of freezer later during the master mix setup). This way, they will site on the benchtop for a shorter time.

Good Practice #14: Always homogenize.

The final reaction volume should be as homogenous as possible to achieve PCR results consistent with each other and to prevent accidental failed PCRs. Always homogenize either by pipetting or by vortexing. This homogenization issue becomes critical after DNA polymerase is added to the master mix. As the enzyme is stored in glycerol, which is much denser than the rest of the master mix components, the enzyme solution will directly sink to the bottom of the tube. Therefore, thorough homogenization of the master mix is a *must* after the addition of the enzyme.

Good Practice #15: Spin tubes before putting them into the thermocycler.

Solutions on the walls will be exposed to a different temperature than the block will intend to heat them to. It is best to have all the volumes at the bottom of the tubes. A quick spin at a low speed will do the job. This will ensure the consistency between PCRs.

Good Practice #16: Use tight racks.

It is better to have a rack into which PCR tubes fit tightly (i.e., tubes are not sitting lose inside the rack). This way tube handling becomes easier, most accidents of losing the tube and sample in it are prevented, and the opening and closing of the tube caps can be easily done using single hand.

Good Practice #17: Put the tubes into a "preheated" thermocycler.

It is a good practice to put tubes at the moment when the program is not counting back from the initial denaturation step, but when the tube holding plate is already heated to over 70°C. Check how many seconds thermocycler requires to get to 95°C after it is started, and prepare a routine procedure so that the tubes can be put inside the cycler on time (try to manage to quick spin the tubes by that time). If this leads to a more than 30-s delay, it is better to stop the thermocycler and restart the program.

This good practice is particularly advised when one is experiencing the "nonspecific bands" problem, the chances of getting nonspecific amplifications are reduced when the first temperature that the reaction tube experiences is above 70°C. However, most modern PCR machines heat up very quickly and this preheating step becomes relatively unnecessary.

A more stringent way of controlling nonspecific amplifications is to use the hot-start technique (discussed in Chapter 4).

Good Practice #18: Use a hot lid or mineral oil to prevent evaporation.

Almost all modern PCR machines have an option of preheating the lid that touches the caps of the tubes. The lid temperature is usually set to 105°C. This step—having the tops of the tubes hot before the plate that touches the bottom and the actual samples is getting hot—is important to prevent evaporation of water from the reaction volume. If this happens, the reactants' concentrations will increase each cycle. This may inhibit and/or decrease the yield of the reaction. If the PCR machine does not have the "hot-lid" option, add a drop of mineral oil to each tube before putting tubes into the thermocycler. Mineral oil is less dense than the solution, it will float at the top of the solution and form a seal that will prevent water from evaporating.

Good Practice #19: Set up reactions on ice.

I, personally, never set up reactions on ice (I used to do this only during my first year of doing PCR), but it *is* a good practice to set up PCR on ice or a cooling block, it will eliminate problems related to degradation and inactivation of certain PCR components (i.e., dNTP, Taq, primers, and DNA). In case problems, such as degradation of PCR product bands and smears, are faced it is *advised* that an ice or cooling block be used for the PCR setup (see Chapter 4).

Good Practice #20: Put the PCR components in a correct order.

There is no single correct order of putting the PCR components into the master mix. However, it is important to always put dH$_2$O either as the first or second component. Water constitutes the largest volume that is added to the master mix. If two other components are added first, one of them may precipitate and the PCR will be ruined. One may or may not notice the precipitate. Starting with water ensures that every component will be at a significant dilution from its stock concentration and there will be no local high concentrations of some of the components and thus no precipitation of them.

The ending is also almost universal: add dNTP just before the enzyme (they are very labile) and put the enzyme just before aliquoting the master mix.

Good Practice #21: Always use PCR-grade water for reactant dilutions.

There are many suppliers of PCR-grade (DNase-free) water. If a lab performs many PCRs, consider buying bottles as large as 1–2 L and take smaller aliquots. Try to keep PCR water in 1-mL aliquots. It is also beneficial to store water frozen at −20°C (preferably in 0.5-mL aliquots), to prevent any growth of microorganisms that may have contaminated the water.

Good Practice #22: Choose thin-walled PCR tubes.

PCR tubes with thin walls will heat and cool fast, and nonspecific reactions at the intermediate temperatures will be avoided.

Good Practice #23: Always know the template concentration and its purity.

It is truly important to at least have an estimate of the template concentration. If one does not know the concentration of the template, too much or too little of it may be added during PCR. One of the simplest ways to estimate DNA/RNA concentration is to measure the OD_{260}. The concentrations in micrograms per microliter are then calculated from these simple formulas:

$$[\textbf{dsDNA}] = \text{dilution factor} \times OD_{260} \times 50$$

$$[\textbf{RNA}] = \text{dilution factor} \times OD_{260} \times 40$$

$$[\textbf{ssDNA}] = \text{dilution factor} \times OD_{260} \times 33$$

$$[\textbf{oligo}] = \text{dilution factor} \times OD_{260} \times 33$$

Alternatively, visit http://www.sciencelauncher.com/od260.html for a handy online calculator.

The purity of a DNA sample can be deduced by measuring OD_{260} and OD_{280} of the DNA samples and then estimating from following rules:

Pure DNA has an OD_{260}/OD_{280} ratio of 1.8–2.0.

Pure RNA has an OD_{260}/OD_{280} ratio of 1.9–2.1.

For DNA samples:

- If OD_{260}/OD_{280} is smaller than 1.8 → the sample has protein contamination.
- If OD_{260}/OD_{280} is larger than 2.0 → the sample has RNA contamination.

PCR is robust enough to tolerate slight deviations from the template concentration, so OD_{260} estimation is a good enough estimate for this application. However, in some cases, DNA concentration estimation by the usage of a DNA-intercalating fluorescent dye, such as Hoechst 34342 (see Appendix E), might be considered.

Precipitating template DNA by cold ethanol is a good way of getting rid of contaminations:

1. Add 3 volumes of 100% ethanol (that was cooled at −20°C for 0.5–1.0 h; ethanol does not freeze at −20°C) to 1 volume of DNA solution.
2. Invert the tube 5–6 times.
3. Spin the tube at 16,000*g* for 5 min.
4. Carefully discard the supernatant by using pipette tip.
5. Add solvent (dH$_2$O or TE Buffer) to dissolve the DNA pellet.
6. Mix by pipetting.
7. Requantify the DNA concentration and purity.

Note: Isopropanol can be used instead of ethanol.

Note: If the amount of DNA was low, increase the incubation with the alcohol to 2 h at −20°C.

To quickly get rid of vast protein contamination, simply add proteinase K (check the manual/booklet of proteinase K to make sure the amount added is not too much or too less) to the template DNA/RNA solution and then denature the proteinase at 95°C for 10 min before using the template solution in PCR.

Good Practice #24: Always use small tips for small volumes.

It is wise to use small, 1–10 μL, tips for P10 for volumes below 5 μL, as the outer wall of larger tips will introduce large errors to the volumes. Even if the pipette is 100% calibrated, the pipette will definitely take up more volume than intended and this will lead to inconsistencies and, in many cases, failed PCRs.

Good Practice #25: Have and use a personal set of reagents.

Usually all of the reagents, except for the enzyme, are in vast numbers so that each member of a lab can get a separate tube of the enzyme buffer and $MgCl_2$. This is because each package of the enzyme comes with an excess of buffer and $MgCl_2$ solutions. Label the buffer and $MgCl_2$ tubes and only use them for three reasons: (1) others may not be cautious enough and may contaminate their PCR reagents, (2) you may accidentally contaminate someone's reagent and someone's research may be affected, and (3) experiments will be more consistent if the same tube of a reagent is used. If a common dNTP solution is not present, just prepare a personal large-volume aliquot (0.2–1 mL) and several small aliquots (25–50 µL) and keep them frozen at −20°C.

Good Practice #26: Try to use filtered pipette tips.

Filtered pipette tips are tips that have a white air-conducting material that does not allow the liquid to pass into the pipette, thus preventing it from contamination. Also, these tips prevent the samples from being contaminated by a dirty pipette. Not all labs use these tips because of their higher price. But if available, use them for pipetting PCR reagents, especially template solutions. Usage of filtered pipette tips will prevent any contamination and especially cross-contamination between samples. Reagents, such as dH_2O, buffer, $MgCl_2$, primers, dNTPs, and enzymes usually do not contaminate the pipettes in a way that directly harms PCR results, but it is better to handle PCR reagents with filtered pipette tips, to prevent a somehow-contaminated pipette from contaminating the reagents. Normal pipette tips can be used for post-PCR steps.

Good Practice #27: Label PCR tubes with enough detail.

Try to write enough details about each PCR on the caps or sides of tubes, so that the same samples can be reused (if there is enough volume to reuse). Usually putting a date on one tube or the master mix and using color pens helps a lot. If recorded in a notebook, it helps in checking which PCR was performed on that particular date and with that particular color. Labeling racks also help in some cases.

Good Practice #28: Always keep a record of PCRs performed.

Whether PCR tubes are labelled in detail or not, it is best to have a logbook dedicated to PCR or a regular lab notebook in which PCR conditions are written down. The idea is to be able to reproduce a PCR even 2 years later and in case of optimizing a PCR, to see what conditions were changed and what results were obtained. Keeping a PCR logfile on computer is a better idea; copy–paste will help a lot, as usually most of the conditions are repeated and only parameters, such as template, primers, and annealing temperature are changed. The photos of the agarose gels can be printed or simply pasted into a computer file (e.g., Word file).

Good Practice #29: Do not freeze–thaw the template too much.

Template DNA quality decreases after each freeze–thaw cycle. Therefore, this number needs to be kept as low as possible. If a large volume (>100 μL) of a template solution is present and no more than 2 μL are used per PCR, it is better to take out a working aliquot of about 25–30 μL and keep it at +4°C, if it is used frequently (everyday) or at −20°C if it is not used so frequently (1–2 times a week). Then, freeze the rest of the template solution at −20°C or −80°C and again take 25–30 μL aliquots whenever the working aliquot finishes.

Good Practice #30: Prevent dA overhangs from degradation.

The Taq-generated dA overhangs, which are used in certain cloning procedures, tend to degrade over time. If PCR products are to be cloned, make sure that they are either used right away or frozen at −20°C (or even better at −80°C). Also, make sure that the samples are not freeze–thawed too many times.

Good Practice #31: Always run a DNA ladder or a previously checked PCR.

It is a good practice to run a positive control, which should give a strong signal on agarose gel, each time a PCR is checked. This way, if no bands are visualized on the gel, one can troubleshoot whether the detection system that has failed or the PCR did not work at all. A molecular weight ladder is the best positive control because the molecular weight of the amplicons can be estimated. A PCR sample that was previously checked on agarose gel would work fine too.

Good Practice #32: Use appropriate agarose gel concentrations for checking PCRs.

Agarose %	DNA sizes (kb)
0.6	1–20
0.7	0.8–10
0.9	0.5–7.0
1.2	0.4–6.0
1.5	0.2–3.0
2	0.1–2.0

Short fragments should be checked on high-percentage agarose gels, so that they can be resolved properly. In other words, it is very hard if not impossible to separate 100 bp from 120 bp at a 1% gel, and it becomes difficult to determine the correct size. If a molecular weight ladder is used (which is recommended), the shorter fragments of the ladder will not resolve at low-percentage gels. Long fragments will not be separates from each other either; they need a low-resolution gel (i.e., low-percentage gel) to be resolved from each other. Thus, use the values in this table to choose which concentration of agarose gel should be prepared for each PCR. 1.5% agarose gels are one of the most commonly prepared ones. This is because they can handle a wide range of double-stranded DNA fragments, from 200 bp to 3 kb and most of the PCR targets lay within this range.

Good Practice #33: Pipette volumes according to the pipettes' volume ranges.

A volume taken outside of a pipette's range will not be exact. Do not try to pipette 10 μL with a P200 (20–200 μL). The rule of thumb for the lower limit is that one should not pipette a volume that is below the 10% of the pipette's capacity. For instance, the lower limit for a P200 is 20 μL and for a P10 it is 1 μL. The upper limits are usually the capacities of the pipettes (Never go beyond the maximum capacity! You will break the pipette!), but it is more accurate to pipette two small volumes instead of one bigger one and at the upper limit of the pipette. For example, if 190 μL needs to be pipetted, it is easier and more accurate to add 90 + 100 μL instead of going up to 190 μL.

Summary of good PCR practices:
1. Have the calculations done before putting one gloves for PCR.
2. Always prepare a master mix for *n* + 1 samples.
3. Check if all the necessary reagents are present before starting.
4. Use dedicated pipettes.
5. Use calibrated pipettes.
6. Work in a dedicated workspace with a dedicated lab coat.
7. Do not immerse pipette tips too deep into solutions.
8. Always thaw–vortex–spin the $MgCl_2$ solution.
9. Always use put on new gloves before starting a PCR setup.
10. Wipe the benchtop with EtOH.
11. Make sure the racks are clean.
12. Make sure the tubes are clean.
13. Store dNTP in small aliquots and thaw them at the end.
14. Always homogenize.
15. Spin tubes before putting them into the thermocycler.
16. Use tight racks.
17. Put the tubes into a "preheated" thermocycler.

18. Use hot lid or mineral oil to prevent evaporation.

19. Set up reactions on ice.

20. Put the PCR components in a correct order.

21. Always use PCR-grade water for reactant dilutions.

22. Choose thin-walled PCR tubes.

23. Always know the template concentration and its purity.

24. Always use small tips for small volumes.

25. Have and use a personal set of reagents.

26. Try to use filtered pipette tips.

27. Label PCR tubes with enough detail.

28. Always keep a record of PCRs performed.

29. Do not freeze–thaw the template too much.

30. Prevent dA overhangs from degradation.

31. Always run a ladder or a previously checked PCR.

32. Use appropriate agarose gel concentrations for checking PCRs.

33. Pipette volumes according to the pipettes' volume ranges.

CHAPTER 4

Optimization and Troubleshooting

Abstract

This chapter discusses various polymerase chain reaction (PCR) optimization strategies and describes how to troubleshoot specific problems based on the outcome of PCR. The optimization session touches on the most critical optimization variables of a PCR reaction, including annealing temperature, Mg^{++} concentration, and ways to increase the precision of a reaction (hot-start and touchdown PCR). Various additives and their role in altering certain parameters of PCR are described. The troubleshooting session identifies and gives solutions to specific problems in a PCR reaction, including failure to amplify a desired product, generation of multiple undesired products, smears observed in gels, low-yield PCR products, low fidelity, and other potential problems, such as primer dimers and fidelity of a reaction.

Creativity can solve almost any problem. The creative act, the defeat of habit by originality, overcomes everything.

(George Lois)

What work I have done I have done because it has been play. If it had been work I shouldn't have done it. Cursed is the man who has found some other man's work and cannot lose it. When we talk about the great workers of the world we really mean the great players of the world.

(Mark Twain)

If all the good practices discussed in the preceding chapter were followed and still the result of the polymerase chain reaction (PCR) is not satisfactory, then what needs to be done is the optimization of the PCR conditions, that is, troubleshooting. This chapter presents the rules for PCR optimization, common optimization strategies, and a problem-specific troubleshooting guide.

4.1 OPTIMIZATION RULES

Optimization is the actual "research" part of performing PCR. In other words, PCR is so simple that the only real brainstorming is done when PCR does not work properly. However, no matter how easy or difficult the problem is, here is a set of rules that we advise to be followed for successful optimization and troubleshooting. We will use the terms "optimization" and "troubleshooting" interchangeably, as troubleshooting a problem related to PCR is essentially optimizing the PCR conditions and vice versa.

Rule #1: Change only one component at a time.

Rule #2: Write down which parameter was altered and how it was altered.

Rule #3: Save all the gel images that were recorded throughout the optimization.

4.2 GENERAL OPTIMIZATION STRATEGIES

4.2.1 Annealing Temperature Gradient

Before gradient thermocyclers were invented, people had to spend days to optimize their PCRs' annealing temperature parameter. Now, with such a wonderful tool, optimization of annealing temperature is only a matter of a single PCR. Given that a PCR protocol has previously worked with a current type of template (gDNA or plasmid) and a current PCR system (PCR machine, buffers, and enzyme), in theory, the only parameter that needs optimization for each new primer set is the annealing temperature (Fig. 4.1).

Even though the main role of Mg^{++} ions is to serve as a cofactor for DNA polymerase, Mg^{++} ions bind to deoxynucleotidetriphosphates (dNTPs), primers, and template DNA too. A fraction of Mg^{++} ions that is added to the reaction will always be chelated by these PCR components. Therefore, whenever the concentration of any of these components is altered, in the case if the PCR result is not satisfactory, the first thing that has to be optimized is the Mg^{++}.

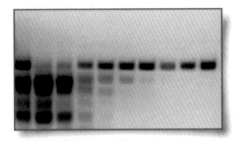

Figure 4.1 *Annealing temperature gradient.* Decreasing the annealing temperature too low leads to nonspecific amplification *(lane 1)* and raising it too high leads to lack of any amplification.

4.2.2 Mg^{++} Titration

As it is easy and inexpensive to set up, Mg^{++} titration is one of the most widely used optimization/troubleshooting approaches. It is also one of the first approaches that comes to mind if one gets a severely negative result (smear or too many nonspecific bands).

Before starting a Mg^{++} titration, check whether the buffer has Mg^{++} already (companies usually supply buffers that are Mg^{++} free, but some include a magnesium salt at a certain concentration). If the buffer already has Mg^{++} in it, during the titration, the concentration of Mg^{++} ions inside the buffer has to accounted too. One can also switch to a Mg^{++}-free buffer of the same supplier.

Following the Rule #1 of all optimization procedures, which states to change only one parameter at a time, keep all the other parameters identical to the previous PCR that had a problem with its results (nonspecific amplification, smear, or no bands) and change only the final Mg^{++} ion concentration.

The usual range for Mg^{++} titration is 0.5–3.5 mM. The titration points can be set in either 0.5-mM or 1-mM increments. In other words, the titration can be fine (0.5, 1.0, 1.5, 2.0, 2.5, 3.0, 3.5, and 4.0 mM) or it can be coarse (1, 2, 3, and 4 mM). The fine titration will give a better estimate of the optimal Mg^{++} concentration.

If all the other parameters, other than Mg^{++} concentration, are optimal, one should get a very good result in at least one of the titration points. If this does not happen, return to the suggested starting point for Mg^{++} concentration (1.0–1.5 mM) and this time try to use one of the other troubleshooting strategies (Fig. 4.2).

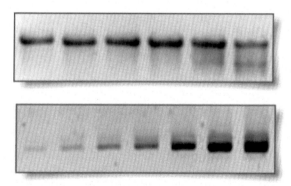

Figure 4.2 *Mg^{++} concentration gradient.* Two experiments were done with two different primer sets. Mg^{++} concentration is increased by 0.5 mM, starting at 4 mM. The low yield at low concentrations and low specificity at high concentrations is evident for both primers sets.

Usually, 1.5 mM $MgCl_2$ is optimal.

Mg^{++} concentration is affected by the amount of template DNA, primers, and dNTPs.

If the Mg^{++} concentration is lower than the optimum, Taq DNA polymerase is not working at its full capacity and as a result, the yield is low (bands are faint). Conversely, excess free magnesium reduces enzyme fidelity and increases the level of nonspecific amplification. For these reasons, it is important to empirically determine the optimal $MgCl_2$ concentration for each reaction.

> Mg^{++} *titration*: Add 1.0, 1.5, 2.0, 2.5, and 3 μL of a 25 mM $MgCl_2$ stock to generate Mg^{++} concentration gradient.

Except for cases of the "special case PCRs," extraordinary primer sets, and unusual templates, it can be said that *optimum PCR conditions* for almost any primer pair cannot escape this "net" produced by a scan of Mg^{++} titration and temperature gradient. In other words, for any given regular primer set, there will be at least one point in the "net" produced by Mg^{++} titration and temperature gradient, where PCR quality is very good (Fig. 4.3).

However, doing this whole 8- × 12-point Mg^{++} titration–temperature gradient "net" is wasteful and unnecessary. As it is usually not necessary to employ all the 12 columns for temperature gradient and an 8-point Mg^{++}

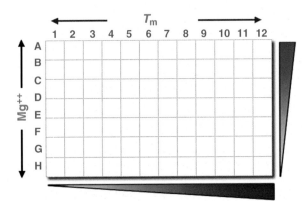

Figure 4.3 *The theoretical version of Mg^{++} titration–temperature gradient "net."* The temperature gradient is obtained by running the thermocycler at an annealing temperature gradient mode and Mg^{++} titration is done by adding increasing volumes of $MgCl_2$ and corresponding decreasing volumes of water, for each sample set (i.e., each row will have the same Mg^{++} concentration). At least one of the squares will be the optimum polymerase chain reaction (PCR) condition, even for the most difficult primer pair.

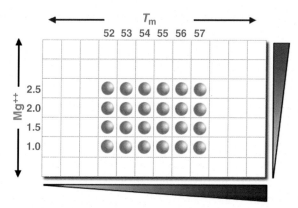

Figure 4.4 *The feasible version of Mg⁺⁺ titration–temperature gradient "net."* The recommended settings for this net are: Mg⁺⁺ titration in the 1–2.5 mM range and annealing temperature gradient in the 52–57°C range with 1°C increments, assuming that 55°C is the T_m−4 value for the primer set (i.e., the calculated average T_m for the primers set is 59°C).

titration is not required at all, perhaps a smaller version of this net can be done (Fig. 4.4).

4.2.3 Hot Start

Hot start is another technique for avoiding nonspecific amplifications, including primer dimers. The idea here is that a missing critical component is added (the enzyme, dNTPs, or $MgCl_2$) *after* the samples are heated up to about 80–85°C. This prevents nonspecific amplifications that would occur if the samples contained all the necessary components from the beginning and were heated up slowly, thus allowing priming at wrong sites and amplification of nonspecific sequences. The classical way of doing this was waiting until the samples were heated up to 80–85°C, opening the tubes, and adding the missing critical component. However, this procedure is tedious and introduces the risk of contaminating the samples.

Several "hot-start" techniques have been developed to avoid this tedious classical procedure:

- A monoclonal antibody that binds to the active site of the enzyme and prevents it from being active until the temperature of the sample reaches 90–95°C, when the antibody denatures and dissociates from the enzyme (e.g., PfuTurbo Hotstart DNA polymerase from Stratagene, Platinum Taq DNA polymerase from Invitrogen, and TaqStart from Clontech).
- A chemically modified DNA polymerase that becomes active only after it has been exposed to 95°C for about 5–10 min (e.g., AmpliTaq Gold from Applied Biosytems).

- Wax beads that contain the missing critical component and that melt again only when the temperature of the sample reaches 75–80°C.

4.2.4 Touchdown PCR

If a situation arises where the nonspecific amplification cannot be controlled by changing the annealing temperature and/or Mg^{++} ion concentration, one can use a technique called "touchdown PCR." The principle is that the annealing temperature at the first 5–10 cycles of the PCR is set 3–10°C higher than the theoretical T_m of the primers and then it is decreased to the T_m of the primers for the remaining 20–30 cycles. The idea here is that the first 5–10 cycles are done at high-stringency conditions and at these higher temperatures, only very few specific products will be synthesized and non-specific products will not be synthesized at all. As the cycles pass on, the amount of specific amplicons accumulate and they serve as templates for further cycles. By the time the annealing temperature is lowered to the degree of T_m, there will be a significant amount of specific amplicons, usage of which as templates will overcome nonspecific amplifications.

Touchdown PCR can be used for various problems with PCR. It even compensates for the suboptimal concentrations of $MgCl_2$ and dNTPs. Also, problems due to primer–template mismatch can be tackled by using touchdown PCR.

If the thermocycler's programming allows so, one can program the machine to decrease the annealing temperature by 0.5°C per cycle, which will give a gradual decrease to the T_m. If this kind of programming is not possible because of the PCR machine's old software, series of cycles with decreasing annealing temperature will have to be entered manually.

4.2.5 Additives

With nucleic acids of high-GC content, it may be necessary to use harsher denaturation conditions.

Additive	Final concentration[a]	Notes
DMSO	2–10%	Secondary structure destabilizer, use with high-GC% targets, DMSO often helps in amplifying products of >1 kb
Glycerol	1–10%	Improves the amplification of high-GC% templates, improves thermal stability of the enzyme

Additive	Final concentration[a]	Notes
BSA	10–100 µg/mL	Enzyme stabilizer, enzyme inhibitor binding
Formamide	1–10%	Greatly increases the specificity of PCR
PEG 6000	5–15%	Useful when DNA template concentration is very low
Tween 20	0.05–1%	Increases yield, may increases nonspecific amplification, enzyme stabilizer, secondary structure destabilizer
Triton X-100	0.01–1%	Increases yield, may increases nonspecific amplification, enzyme stabilizer, secondary structure destabilizer
Nonidet P40	0.1–1%	Increases yield, may increases nonspecific amplification, enzyme stabilizer, secondary structure destabilizer
Tetramethylammonium bromide	5%	Enzyme stabilizer
Tetramethylammonium chloride	15–100 mM	Eliminates nonspecific amplifications
Betaine[b]	1–1.7 M	Secondary structure destabilizer, use with high GC% targets

BSA, Bovine serum albumin; DMSO, dimethyl sulfoxide; PEG, polyethylene glycol.
[a]Recommended ranges.
[b]Also known as trimethylglycine.

Note: When stored at room temperature, dimethyl sulfoxide (DMSO) tends to destroy the walls of plastic tubes. It is best to aliquot DMSO into small volumes and store at −20°C.

Note: When using DMSO at concentrations greater than 5%, the annealing temperature in the PCR program must be decreased because in the presence of DMSO, the T_m of primers decreases. If DMSO is used at a final concentration of 10%, the annealing temperature should be decreased by 6°C.

The best approach in using an additive, to troubleshoot a problem or improve PCR quality, is to make a titration of several concentrations within the recommended concentration range (as listed in the Table earlier). For instance, if high GC% is a problem, then try adding DMSO to the PCR reaction, make 2, 4, 6, and 8% (final DMSO concentration) titration points and check which one gives best results for the target of interest.

4.3 TROUBLESHOOTING SPECIFIC PROBLEMS

4.3.1 Problem Type 1

No PCR products were observed.

PCR reaction components are described in Table 4.1:
Cycling conditions are described in Table 4.2:
Accidental problems:

- Wrong template: The target sequence is not present in the template.
- Thermocycler stopped due to power failure.
- Cycling conditions were programmed incorrectly.

4.3.2 Problem Type 2

Multiple PCR products were observed.

PCR reaction components are described in Table 4.3:
Cycling conditions are described in Table 4.4:
Nonspecific products longer than the target:

- Decrease annealing time.
- Increase annealing temperature.
- Decrease extension time.
- Decrease extension temperature to 62–68°C.
- Increase buffer concentration to ~1.5×.
  Nonspecific products shorter than the target:
- Increase annealing temperature.
- Increase annealing time.
- Increase extension time.
- Increase extension temperature to 74–78°C.
- Decrease buffer concentration to ~0.8×.

4.3.3 Problem Type 3

Smears were observed.

PCR reaction components are described in Table 4.5:
Cycling conditions are described in Table 4.6:

4.3.4 Problem Type 4

Observed bands were faint (low yield).

PCR reaction components are described in Table 4.7:
Cycling conditions are described in Table 4.8:

4.3.5 Problem Type 5

Bands observed in controls.

Table 4.1 Potential problems and solutions with regard to PCR reaction components in the case of no PCR amplification

Potential problems	Potential solutions
Wrong buffer was used	Some enzymes require Triton X-100 in the reaction at a final concentration of 0.1%; check the manual/booklet of the enzyme for this information and verify that the correct buffer is being used
Primers were not adequately designed	Refer to Chapter 2 and verify that the primer has all the necessary properties
Primer concentration was too low	Usually the final concentration is about 0.4–1.0 µM; check the calculations and redilute the primers if necessary
Mg^{++} concentration was too low	Perform a Mg^{++} titration
Mg^{++} solution was not vortexed after thawing	Always vortex the $MgCl_2$ solution after thawing
dNTP concentration was too low	If the target is longer than 30 kb, increase the dNTP concentration to about 0.5 mM; make a dNTP titration; keep in mind that more Mg^{++} ions will be chelated by the increased dNTP
Template concentration was too low	(1) Too much plasmid DNA was added or (2) not enough genomic DNA was added; final concentrations ~10 ng/µL are suggested
Template concentration was too high	Do not input too much template; it may inhibit the reaction completely; stay below 10 ng/µL as the final concentration
Template is of low quality (degraded)	Check template quality by agarose gel electrophoresis; for genomic DNA, run a 0.8% gel
Template is of low quality (not pure enough)	Check OD_{260}/OD_{280} value of the template; see Chapter 3; repurify the template or ethanol precipitate it and resuspend in an adequate volume of dH_2O
Template has a high-GC content and/or forms stable secondary structures	Try using an appropriate additive
One of the components was old/degraded and/or contaminated	Thaw a new dH_2O, buffer, and $MgCl_2$ vial; use a fresh dNTP solution, a new dilution of primers or another enzyme of the same kind and brand; (always change only one component at a time)

Table 4.2 Potential problems and solutions with regard to cycling conditions in the case of no PCR amplification

Potential problems	Potential solutions
Denaturation temperature was too low/high	Usually 95°C is sufficient; but the template might have unusual properties, such as high-GC content; in this case, increase the denaturation temperature up to 98°C in 1°C increments
Denaturation time was too short/long	Usually 30 s is sufficient; if it is known that the DNA has a high-GC content, increase the time up to 45 s or even 1 min
Annealing temperature was too high	Make an annealing temperature gradient in the lower range of the primers' T_m values
Extension time was not long enough	The rule of thumb is to have 45 s per 1 kb of the target; adjust the number of minutes accordingly
Too much evaporation	Use a hot lid (105°C) or add mineral oil to the samples to prevent evaporation

Table 4.3 Potential problems and solutions with regard to PCR reaction components in the case of multiple PCR products

Potential problems	Potential solutions
Wrong buffer was used	Some enzymes require Triton X-100 to be present in the reaction at a final concentration of 0.1%; check the manual/booklet of the enzyme for this information and verify that the correct buffer is used
Primers were not adequately designed	Refer to Chapter 2 and verify that the primer has all the necessary properties
Primer concentration was too high	Decrease primer concentration by 0.5 µM; usually a range of 0.4–1.0 µM is optimal
Mg^{++} concentration was too high	Perform a Mg^{++} titration
Mg^{++} solution was not vortexed after thawing	Always vortex the $MgCl_2$ solution after thawing
dNTP concentration was too high	Usually 0.2–0.25 mM is sufficient for most targets shorter than 35 kb; decrease in 0.05-mM increments
Template has a high-GC content and/or forms stable secondary structures	Try using an appropriate additive
Template has a high-GC content and/or forms stable secondary structures	Try using an appropriate additive, such as 5% DMSO

(Continued)

Table 4.3 Potential problems and solutions with regard to PCR reaction components in the case of multiple PCR products (*cont.*)

Potential problems	Potential solutions
Template concentration was too high	Stay below 10 ng/μL as the final template concentration
Enzyme concentration was too high	Check the recommended concentration of the enzyme from the enzyme supplier's manual/booklet

Table 4.4 Potential problems and solutions with regard to cycling conditions in the case of multiple PCR products

Potential problems	Potential solutions
Annealing temperature too low	Make an annealing temperature gradient in the higher range of the primers' T_m values
Too many cycles	Decrease the number of cycles by 3–5; try to keep the cycle number below 35
Touchdown PCR needed	See Section 4.2.4
Hot start needed	See Section 4.2.3

Table 4.5 Potential problems and solutions with regard to PCR reaction components in the case of smearing

Potential problems	Potential solutions
Mg^{++} concentration was too low	Perform a Mg^{++} titration
Mg^{++} solution was not vortexed after thawing	Always vortex the $MgCl_2$ solution after thawing
Template concentration was too high	Stay below 10 ng/μL as the final template concentration
Template is of low quality (degraded)	Check template quality by agarose gel electrophoresis; for genomic DNA, run a 0.8% gel
Template is of low quality (not pure enough)	Check OD_{260}/OD_{280} value of the template; see Chapter 3; repurify the template or ethanol precipitate it and resuspend in an adequate volume of dH_2O
dNTP concentration not enough	Make a dNTP titration with 0.5-mM increments
Enzyme concentration was too high	Check the recommended concentration of the enzyme from the enzyme supplier's manual/booklet
Proteases/nucleases in the samples	Incubate samples at 95°C for 10 min before adding the enzyme; set up reactions on ice

Table 4.6 Potential problems and solutions with regard to cycling conditions in the case of smearing

Potential problems	Potential solutions
Too many cycles	Reduce the cycle number by 3–5 cycles
Denaturation temperature too low	Increase the denaturation temperature in 1°C increments
Extension time too long	Decrease the extension time in 1-min increments
Touchdown PCR needed	See Section 4.2.4

Table 4.7 Potential problems and solutions with regard to PCR reaction components in the case of the appearance of faint bands

Potential problems	Potential solutions
Primer concentration was too low	Usually the final concentration is about 0.4–1.0 μM; check the calculations and redilute primers if necessary
Mg^{++} concentration was too low	Perform a $MgCl_2$ titration
Mg^{++} solution was not vortexed after thawing	Always vortex the $MgCl_2$ solution after thawing
Template concentration was too low	Final concentrations ~10 ng/μL are suggested
Template has a high-GC content and/or forms stable secondary structures	Try using an appropriate additive, such as 5% DMSO
Enzyme concentration was too low	Check the recommended concentration of the enzyme from the enzyme supplier's manual/booklet

Table 4.8 Potential problems and solutions with regard to cycling conditions in the case of the appearance of faint bands

Potential problems	Potential solutions
Annealing temperature was too high	Make an annealing temperature gradient in the lower range of the primers' T_m values
Extension temperature was too high	Decrease extension temperature in 1°C increments
Extension time was too low	Increase extension time according to the rule of thumb: 45 s per 1 kb of target sequence
Not enough number of cycles	The cycle number should usually be higher than 20 but lower than 35; increase the cycle number by 5

If any band (even a faint one) is observed from the negative ("no DNA") control tube, this may mean one of the following:

1. *There was a mistake during loading*: Load the negative control again and verify that there is a band.
2. *There was an overflow from a neighbor well (if the band is very faint)*: Leave a blank well (do not load anything) between the negative control and the first sample.
3. *There is a contamination in at least one of the pipettes*: Send them for cleaning.
4. *There is a contamination in at least one of the reagents*: Change the cheap reagents right away (working tubes of dH_2O, primers, buffer, or $MgCl_2$). dNTP is not so inexpensive, but discard it too, if there is not much volume left in the working dNTP dilution.
5. *Contamination was accidentally introduced during PCR setup*: Set up the PCR again.

One can decide which of the five possibilities happened in a particular case. Use your best judgment to take action, but changing reagents (case "4") is usually the case, and it is the easiest and quickest way to assess where the problem is.

4.3.6 Other Problem Types

Primer dimers are observed:
- Annealing temperature is suboptimal.
- The $3'$-ends of the primers are complementary.
- Primer concentration is too high.
- Target template concentration is too low.
- Primers need to be longer.

Fidelity is low:
- Decrease dNTP concentration to 0.1 mM and then reoptimize Mg^{++} concentration.
- Set fewer cycles.
- Use high-fidelity enzyme (e.g., Pfu).

CHAPTER 5

Tips and Tricks

Abstract

In this chapter, 13 tips and tricks that are a guide to save time and effort, as well as employ good practices when performing polymerase chain reaction (PCR) reactions either in large or small numbers are laid out. Practical tips and tricks for efficient, precise, and successful PCR reactions include preparing and storing premix solution for use in routine PCR reactions, modifying cycling parameters to shorten the PCR time for primers with high T_m, ways to thaw solutions quickly, dealing with nonspecific amplifications, as well as good practice techniques, such as removing bubbles prior to reaction and delivering the premix solution correctly and efficiently.

Work joyfully and peacefully, knowing that right thoughts and right efforts inevitably bring about right results.

(James Allen)

Trust yourself, you know more than you think you do.

(Benjamin Spock)

The secret to success is to know something nobody else knows.

(Aristotle Onassis)

#1: If one is performing the same optimized polymerase chain reaction (PCR) frequently, assembling a large-volume (0.5–1.0 mL or even more) premix will save time and effort:

1. Mix dH_2O, buffer, $MgCl_2$, and primers at the optimized ratios.
2. Aliquot this premix at appropriate volumes.
3. Store at −20°C or −80°C.

With such as premix, thawing will only be required when ones needs to set up another round of the routine PCR. Add the enzyme and deoxynucleotidetriphosphates (dNTPs) to it, distribute to PCR tubes, and add the template DNAs to each tube. This premix trick truly saves a lot of time and effort.

Premixes can be set up in various combinations. One can premix only dH_2O, buffer, and $MgCl_2$ or only dH_2O and buffer. But never premix the enzyme because it will lose its activity when frozen without glycerol (enzymes are usually frozen in a buffer that has 50% glycerol).

Make sure that the calculations are correct!

#2: If PCR for only two to three samples is to be performed, do not prepare a master mix for two to three samples. Instead, make a duplicate of each and prepare the master mix for four to six samples. This way one can avoid working with too small volumes that usually introduce pipetting errors and are just hard to work with, and get twice as more material. Increasing the reaction volume is also an alternative, but volumes should be kept small, so that the heating and cooling of the sample is as uniform as possible.

#3: PCR can be done in 40 min! If the target sequence is shorter than 500 bp, cycling parameters can be modified to the following and checked if they work for the primers:

1) 98°C for 30 s
2) 92°C for 1 s
3) 70°C for 20 s
Repeat steps 2–3 for 35 cycles.

The total cycling time should be about 35 min! The annealing and extension steps are fused here. In this case, the primers should have a T_m around 65°C. If this does not work, decrease the temperature of step 3 to 68°C. Also, there is no final 5-min extension time at 72°C. This protocol would work best with fragments smaller than 250 bp, can go as high as 500 bp. Of course, the extension length can be increased for optimization. However, this protocol may or may not work for every primer set. Such two-step PCR can also be performed with the extension temperature closer to the annealing temperature.

#4: If bubbles are present, centrifuge! A quick spin at a low speed should get rid of the bubbles.

#5: dH_2O, buffer, $MgCl_2$, and even primers can be stored at +4°C for a short time, if they are going to be frequently used, and this saves the thawing time. However, do not keep them at +4°C too long and try to freeze overnight. Solutions may become contaminated with some bacteria and that may cause a lot of trouble for PCR. Primers will not be stable at +4°C for very long times too. But they can be safely thawed at the beginning of a week and then frozen only after 2–3 days.

#6: If PCR-grade water is not available, irradiating the water with a UV lamp for 5 mins can yield relatively good quality water for PCR. It is advised to purchase PCR-grade water, though.

#7: Thaw $MgCl_2$ solutions very slowly. The tubes can be put into a heat block (ones with temperature lower than 95°C are preferred); $MgCl_2$ solutions are not heat labile. But, do not let the solution to heat up. Take it out as soon as most of the ice has melted so that the remaining ice will thaw and cool down the solution. It is better to keep all the reagents cool before they are added to the master mix. The same trick can be done with dH_2O, but not with any other reagent (even buffer may be affected if this is done too many times)!

#8: If PCR tubes do not sit tightly in the thermocycler's heat block, a drop of mineral oil can be added for better heat transfer between the block and the tubes. But first, make sure that this practice is approved by other members of the lab.

#9: Beside all the troubleshooting strategies described in Chapter 4, there is a shortcut to save time if there is a problem with nonspecific amplifications. Instead of investing hours or even days trying to optimize the PCR conditions, just cut the band of interest out of the gel (preferably low-melting agarose gel), then purify it with a commercially available kit, and use it as template for a second round of PCR. Brilliant results are expected because the only template this time will be the target itself and there will be no secondary sites for nonspecific amplifications.

#10: The same pipette tip can be used to deliver the master mix to multiple tubes (put the drop of master mix on one side of the tube). Then, if the same template needs to be added to multiple tubes too, use the same pipette tip to deliver the drops of template on another side of the tube (Fig. 5.1).

#11: A rough estimate of where the expected band should run on the agarose gel can be made with the use of bromophenol blue (BPB) and xylene cyanol (XC), which can be used as molecular weight markers. The relative migration rates for these dyes and the double-stranded DNA (dsDNA) amplicons are:

Agarose (%)	BPB	XC
0.5–1.5	400–500 bp	4000–5000 bp
2.0–3.0	100 bp	750 bp
4.0–5.0	25 bp	125 bp

BPB, Bromophenol blue; XC, xylene cyanol.

In other words, in a 2% agarose gel, the BPB front migrates as fast as a 100-bp dsDNA fragment and the XC front migrates as fast as a 750-bp dsDNA fragment.

#12: Use the same thermocycler and even the same block of a thermocycler for a previously optimized PCR. Different thermocyclers of even the same brand and even different blocks of the same machine tend to have slight differences that may lead to suboptimal amplification quality. This is especially true if the machine is old.

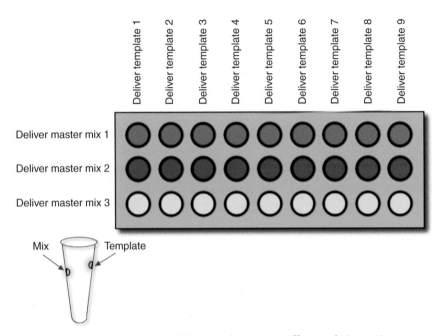

Figure 5.1 *How to make a quick delivery and save time, effort, and pipette tips.*

#13: A table similar to one shown further is a handy tool for preparing routine master mixes of optimized PCRs for different sample numbers. The proportions for one sample are given as "1×" and depending on the number of samples to be prepared, one just needs to look at the microliters in the "(n + 1)×" column and prepare the master mix. The numbers given in the table are real and are optimized for human genomic DNA PCRs.

1×		4×	5×	6×	7×	8×	9×	10×	11×	12×	13×	14×	15×
17	dH$_2$O	68	85	102	119	136	153	170	187	204	221	238	255
2.5	10× Buffer	10	12.5	15	17.5	20	22.5	25	27.5	30	32.5	35	37.5
2	25 mM Mg^{++}	8	10	12	14	16	18	20	22	24	26	28	30
1	10 µM Forward primer	4	5	6	7	8	9	10	11	12	13	14	15
1	10 µM Reverse primer	4	5	6	7	8	9	10	11	12	13	14	15
0.5	12.5 mM dNTP	2	2.5	3	3.5	4	4.5	5	5.5	6	6.5	7	7.5
0.1	5 U/µL Taq	0.4	0.5	0.6	0.7	0.8	0.9	1	1.1	1.2	1.3	1.4	1.5
1 µL per sample	100 ng/µL DNA												

dNTPs, Deoxynucleotidetriphosphates.

CHAPTER 6

Special Cases

Abstract

This chapter provides various tricks, tips, and troubleshooting solutions regarding non-standard polymerase chain reaction (PCR) cases, such as "long PCR," randomly amplified of polymorphic DNA PCR, and quantitative PCR. Optimization information is given on how to amplify very long PCR products, as well as how to amplify DNA targets for which the DNA sequence might be unknown (randomly amplified of polymorphic DNA PCR). In addition, quantitative PCR is described and a guide for detecting potential problems and troubleshooting is provided.

Aerodynamically, the bumble bee shouldn't be able to fly, but the bumble bee doesn't know it so it goes on flying anyway.

(Mary Kay Ash)

Somewhere, something incredible is waiting to be known.

(Carl Sagan)

"Special cases" is the chapter where we discuss the nonstandard polymerase chain reaction (PCR) protocols. Most of the PCR applications are limited to simple amplification of regular short (<1 kb) DNA sequences.

- *Long PCR*

  Length of the target sequences for most PCRs does not exceed 1 kb. However for certain applications, researchers might need to amplify fragments that are longer than 5 kb or even 15 kb. For these applications, the standard PCR protocols have to be slightly modified and in most cases special DNA polymerases have to be used to ensure a high-yield amplification without mutations introduced by PCR. As standard PCR protocol is being significantly modified when dealing with long targets, the new PCR procedure is often called "long PCR."

  Modifications that are made to the standard PCR protocol are:
- The extension time should be 1 min per 1.5 kb. So, if a 9-kb fragment is to be amplified, set the extension time to 6 min.
- For amplifications of targets that are longer than 10 kb, the same rule applies (1 min per 1.5 kb), but after the 10th cycle, gradually increase the extension time (10–20 s per cycle). Almost all of the modern thermocyclers have appropriate programming to do this.

- The extension temperature should be 68–70°C. Try to use enzymes that are best active at these temperatures rather than at 72°C.
- The denaturation step of each cycle should be shorter and at a lower temperature (e.g., 92°C/10 s). This is particularly important for long PCR to avoid extensive loss of enzyme activity required to amplify long targets.
- The primers should be designed so that their T_m is high (65–68°C). With such primers one can even try doing a two-step amplification (annealing and extension steps are fused into a single step at 68°C at the calculated extension time).
- For genomic DNA try to keep primer concentrations within a 0.5–1.0 µM range. For plasmid DNA this range should be 0.2–0.5 µM.
- For targets longer than 10 kb, keep the cycle number below 20.
- Try to use very high-quality template samples.
- *RAPD PCR*
  Randomly amplified polymorphic DNA (RAPD) PCR is a special PCR type where degenerate (nonspecific) primers are used to amplify a DNA target for which the sequence may or may not be known. It is useful for discriminating between different DNA templates and thus can be used to differentiate between various animal species.
- Samples should be loaded with less tracking dye (bromophenol blue and xylene cyanol) to have the estimation of amplicon size at maximum sensitivity.
- For the same reason, gels should be casted without ethidium bromide (EtBr) and stained after the run, by incubating the gels in the running buffer with EtBr, added to a final concentration of 0.1 µg/mL for 10 min in a fume hood.
  With the RAPD PCR process or with PCR with A/T-rich primer/template hybrids, it may be necessary to heat up the sample slowly after annealing. In the example given, the elongation temperature is attained at a speed of 1°C per second.

1. Initial denaturation:	95°C/2 min
2. Denaturation:	95°C/15 s
3. Annealing:	57°C/15 s
4. Elongation:	72°C/30 s (heating rate = 1°C s^{-1})
5. Final Elongation:	72°C/2 min

Repeat steps 2–4 30 times.
- *Quantitative PCR (qPCR) troubleshooting*
  No amplification:
- No target in sample or target below the limit of detection.

- Missing reaction component.
- Incorrectly assigned dye detector.
- Expired reagent.
- Suboptimal PCR conditions (annealing and extension times and cycle number).

  Inconsistent replicates:
- Pipetting errors.
- Master mix not used.
- Low–target copy number.
- Inappropriate cycling conditions.
- Probe degraded or degrading during reaction.
- Addition of too much probe.
- Evaporation.

  Poor efficiency of reactions:
- Poor primer design: Inefficient amplification by the primers.
- Incorrect primer concentration: Primer concentration suboptimal.
- PCR annealing/extension time too short.
- PCR annealing/extension temperature too high or too low.
- $MgCl_2$ concentration suboptimal.
- Low-fluorescent dye intensity.
- Sample inhibition.

  Unexpected signal:
- Instrument calibration (signal detected in wrong channels).
- Contamination (reagent or genomic contamination).

  The standard curve is not linear.
- Exceeded assay sensitivity: Primer concentration needs optimization.
- Exceeded assay capacity: Nucleic acid concentration needs optimization.

  Tips for designing primers for a quantitative PCR assay:
- T_m between 50 and 65°C.
- Avoid repeats of Gs or Cs longer than three bases.
- Place Gs and Cs on the ends of the primers.
- GC content of the primers should be between 50 and 60%.

Glossary and Abbreviations

A
Amplicon A PCR product, PCR band, specific amplified dsDNA sequence.

B
bp   Base pair.
BPB   Bromophenol blue. A loading dye for electrophoresis. Related to XC.
BSA   Bovine serum albumin.

C
cDNA   Complementary DNA. Synthesized from RNA using the reverse transcriptase enzyme.

D
Da   Dalton. Unit of molecular weight.
dA   Deoxyadenine. The nucleotide A that is present in a building block of DNA.
dH$_2$O   Distilled water.
DMSO Dimethyl sulfoxide.
DNases Deoxyribonucleases. A DNA-degrading/cutting enzyme class.
dNTPs Deoxynucleotidetriphosphates. A mix of dATP, dTTP, dCTP, and dGTP.
ds   Double stranded.

E
EtBr   Ethidium bromide.
EtOH   Ethanol.

G
gDNA   Genomic DNA.
GI   Gene ID.

K
kb   Kilobase (1000 bp).
kDa   Kilodalton (1000 Da).

M
MW   Molecular weight (g/mol).
M   Molar (mol/L).
mM   Millimolar (Molar $\times$ 10^{-3}).
μM   Micromolar (Molar $\times$ 10^{-6}).

N
nt   Nucleotides.
NEB   New England Biolabs.

PCR Guru

O

OD	Optical density.
ORF	Open reading frame.

P

PCR	Polymerase chain reaction.
PCR machine	Thermocycler or cycler.
PEG	Polyethylene glycol.

Q

qPCR	Quantitative PCR.

R

RAPD	Randomly amplified polymorphic DNA.
REs	Restriction enzymes.

S

ss	Single stranded.

T

TAE	Tris–acetate–EDTA buffer.
TBE	Tris–borate buffer.
T_m	Temperature (in Celsius) required to denature half of the base pairs of a given DNA duplex. Also called melting temperature. For primers, T_m is the theoretical estimate of the temperature at which the primers are "half dissociated" from the template. For primers that are shorter than 20 nt, the simple formula for T_m estimation is $T_m = 4(G + C) + 2(A + T)$. For longer primers, a "nearest neighbour" calculation is required and thus, computer-based tools should be used.
TNE	Tris–NaCl–EDTA buffer.

X

XC	Xylene cyanol. A loading dye for electrophoresis. Related to BPB.

APPENDIX A

Agarose Gel Electrophoresis Protocol

Materials:
- Agarose
- 1× TAE or 0.5× TBE buffer (see Appendix K)
- 10 mg/mL ethidium bromide (EtBr)
- Flask
- Gel-casting tray
- Microwave
- UV transilluminator
- Samples to be analyzed
- Pipette
- 5× Agarose loading dye (see Appendix K)

Procedure:
1. Calculate how many grams of agarose are needed to prepare the desired percentage agarose gel of desired volume (e.g., for 30 mL of 1% gel, 0.3 g of agarose is required).
2. Assemble the gel-casting tray.
3. Weigh agarose and put it into a flask that is at least 3 times the total liquid volume of the gel to be prepared.
4. Pour the buffer (0.5× TBE or 1× TAE).
5. Boil in a microwave or on a heater–stirrer.
6. Make sure that the solution is transparent.
7. Wait till the flask is cooled down and is touchable.
8. *Go to a fume hood and do all the next steps in the hood.*
9. Add 1.5 μL of 10 mg/mL EtBr per 30 mL (final concentration, 0.5 μg/mL).
10. Swirl the flask.
11. Pour the gel solution into the tray.
12. Keep at room temperature till the gel solidifies.
13. Place the gel into precooled running buffer (0.5× TBE or 1× TAE).
14. Load the wells with samples and molecular weight markers.
15. Run (10–20 min) (1) PCR products: 150 V in 0.5× TBE and (2) plasmids: 100 V in 1× TAE.

16. Analyze the bands through a UV transilluminator.

17. Document the results.

Warning: EtBr is a mutagen and carcinogen! Handle it only in fume hoods, and wear gloves and lab coat. A safer alternative, such as SYBR Green I dye (e.g., SYBR Safe) can be used instead, if very high sensitivity is not required.

Warning: Wear special "anti-UV" goggles. Do not look at the UV source.

Tip: Agarose solutions can be prepared in larger volumes and kept at 50–65°C.

Tip: Use an ELISA microplate (or a sheet of Parafilm) for mixing the loading dye and the samples. Usually 5 μL of a PCR product is sufficient to visualize good bands on an agarose gel.

APPENDIX B

Silver Staining of DNA Gels

Materials:
- Buffers A, B, C, and D
- Gel-staining tray
- dH_2O for rinsing gel
- Transparent file for gel storage
- Tissue papers
- Shaker/rocker

Buffer A (prepare 1 L); Fixative
10% Ethanol
0.5% Acetic acid

Buffer B (prepare 1 L)
0.1% $AgNO_3$
Stir well (!) to dissolve all $AgNO_3$ crystals. Keep in dark.

Buffer C (0.5 L) (prepare fresh!)
7.5 g NaOH
0.05 g $NaBH_4$
Add dH_2O up to 500 mL.
Add 2.4 mL formaldehyde (37%) just before applying to gel.

Buffer D (1L); Development enhancer
7.5 g Na_2CO_3
Add dH_2O up to 1000 mL.

Procedure:
1. Pour 50–100 mL of Buffer A into the staining tray.
2. Separate the gel glasses and cut one corner of the gel for the orientation.
3. Place the gel into the tray.
4. Incubate on shaker for 5 min. While the gel is on the shaker, wash the glasses.
5. Discard Buffer A.
6. Pour Buffer B.
7. Incubate on shaker for 10 min.

8. While the gel is on the rocker prepare the Buffer C.
9. Collect Buffer B for recycling.
10. Rinse the gel in *distilled* water.
11. Pour Buffer C.
12. Incubate on shaker until bands get intense enough. Do not incubate too much to prevent doublet bands from fusing.
13. Discard Buffer C.
14. Pour Buffer D.
15. Incubate for 5 min on shaker to enhance the band intensity.
16. Discard the Buffer D.
17. Take the gel by tissue paper and put onto opened transparent folder.
18. Seal the gel.

APPENDIX C

Plasmid Miniprep Protocol (by using "QIAprep Spin Miniprep" Kit)

Materials:
- Bacterial liquid culture
- Table top centrifuge
- Buffer P1
- Buffer P2
- Buffer N3
- Buffer PB (optional)
- Buffer PE
- Buffer EB or dH$_2$O
- QIAprep columns

Note: All spins with the *columns* are at room temperature.

Procedure:
1. Spin 3–5 mL of bacterial liquid culture (grown for 8–16 h) in a 5-mL tube at 4000*g* for 5 min at 4°C.
2. Discard the supernatant.
3. Add 250 μL of Buffer P1 to the pellet. Resuspend by vortexing.
4. Add 250 μL of Buffer P2. Gently invert the tubes 4–6 times or vortex for 5 s.
5. Add 350 μL of Buffer N3. *Immediately* invert the tubes 4–6 times or vortex for 5 s.
6. Transfer the lysates from 5 mL tube to 1.5-mL tube.
7. *Spin: 16,000g for 5–10 min.*
8. Apply the supernatants to the QIAprep columns (with flow-through collectors).
9. *Spin: 10,000g for 1 min.* Discard the flow through.
10. Optional: Add 0.5 mL of Buffer PB to the column.
11. *Spin: 10,000g for 1 min.* Discard the flow through.

12. Add 0.75 mL of Buffer PE to the column. Let stand for 2–5 min if the miniprep is to be used in salt-sensitive experiments (like blunt-end cloning).
13. *Spin: 10,000g for 1 min.* Discard the flow through.
14. *Spin: 16,000g for 1 min once more.* Discard the flow-through collectors.
15. Put the columns into labeled 1.5-mL Eppendorf tubes.
16. Put 50-μL EB to the center of each column filter.
17. Let stand for 1–3 min (for higher yields, let stand for 3 min).
18. *Spin: 16,000g for 1 min.*
19. Keep the minipreps at +4°C. For long-term storage freeze at −20°C.
20. Make 1% agarose gel. Use 5 μL of the miniprep per well.
21. Measure the OD_{260} and $OD_{260/280}$ ratios.

Trick: Add a clean pipette tip into the 5-mL tube to speed up resuspension of the pellet by vortexing (step 3).

Trick: Prewarm EB or dH_2O in a 1.5-mL tube up to 60–70°C. The net yields of the eluted plasmid DNA will increase.

Trick: Using the vacuum manifold (see Qiagen's website, https://www.qiagen.com/us/shop/lab-basics/qiavac-24-plus/#orderinginformation) instead of centrifugation will greatly speed up the whole process.

Trick: Using a dispenser pipette to deliver the buffers will greatly speed up the whole process.

APPENDIX D

Restriction Enzyme Digestion Protocol

Materials and procedure:
  For a 50-µL reaction, mix:

5 µL	10× Buffer (each enzyme has its own buffer)
5 µL	10× BSA (if not already in the buffer)
1 µL	10 U/µL restriction enzyme
0.5–1 µg	DNA (e.g., plasmid miniprep)
Up to 50 µL	dH$_2$O

BSA, Bovine serum albumin.

1. Thaw all the reagents well.
2. Vortex and briefly spin the tubes.
3. Keep the reagents on ice. Leave the restriction enzyme at −20°C or keep in a tabletop cooler box.
4. Mix the reagents starting from dH$_2$O.
5. Add the restriction enzyme the last.
6. Incubate at 37°C for at least 1 h.
7. To ensure the digestion (or if the enzyme is old) incubate overnight.

Note: Always check the concentration and the expiration date of the restriction enzymes before using them.

Note: The temperature at which restriction enzymes are maximally active varies. Check the datasheet of the enzyme.

Note: The volume of the enzyme should not exceed 10% of the total reaction volume.

Protocol for DNA Concentration Estimation Using Hoechst 33258

Hoechst 33258 (H258):
Bisbenzimide DNA-intercalating fluorescent dye.
Binds to AT-rich regions.
Fluorescence is enhanced under high-ionic strength conditions.
Excitation: 350 nm.
Emission: 450 nm.

Materials:
- dH_2O (preferably filtered through a 0.2–0.4 μm filter)
- 10× TNE Buffer (see Appendix K)
- 1× TE buffer (see Appendix K)
- 1 mg/mL H258 in dH_2O (store in amber bottle at 4°C for up to 6 months)
- 1 mg/mL CalfThymus DNA in 1× TE (store at 4°C for up to 3 months)
- 10 × 10 mm² fluorescence cuvettes

Procedure:
1. Prepare fresh (daily) 2× dye solution: Dilute 20 μL of H258 stock solution (1 mg/mL) with 100 mL 1× TNE. (Store at room temperature in amber bottle.)
2. Prepare standard solutions: Make a threefold serial dilution series of the DNA from 0.02 to 2 μg/mL in 1× TNE.
3. Prepare a blank control: Mix 1 mL of 1× TNE with 1 mL of 2× dye solution.
4. Mix 1 mL of each sample/standard solution with 1 mL of 2× dye solution (the concentration of DNA is halved now for each dilution point).
5. Measure fluorescence of each sample at ~450 nm when excited at ~350 nm.
6. Prepare a standard curve: Fluorescence versus concentration (ng/mL) plot (do not forget to subtract the fluorescence of the blank).
7. Repeat the steps 4 and 5 for each sample, concentration of which you want to learn.

8. Deduce the concentration by using the line equation from the standard curve.

Note: Pipette thoroughly. Do not introduce bubbles.

Note: Hoechst 33258 is a potential carcinogen and mutagen. Always wear lab coat and gloves and work under a fume hood. Wearing goggles is recommended too.

APPENDIX F

Formulas

Rough estimate of melting temperature (T_m):

$$T_{\text{m}} = 2 \times (A + T) + 4 \times (G + C)$$

Where A, T, G, and C represent the number of adenine, thymine, guanine, and cytosine nucleotides, respectively.

Beer–Lambert Law:

$$A = \varepsilon \times c \times l$$

A: Absorbance (unitless).
ε : Extinction coefficient ($M^{-1}cm^{-1}$ or cm^2/mg).
c: Concentration (M or mg/mL).
l: Length of the path through which light passes (cm).

Prediction of PCR yield:

$$\text{PCR yield} = I \times (1 + e)^C$$

I: Amount of target input (mg or pg).
e: Efficiency (%).
C: Cycle number.
For instance, if the efficiency of a particular PCR setup is 80%, then it would take ~24 cycles to obtain 1 mg of a target sequence that was inputted at 1 pg amount.

1 mg = 1 pg × $(1 + 0.8)^C$
1,000,000 = 1.8^C
$10^6 = 1.8^C$
$10^{6/C} = 1.8$
$\log_{10}(1.8) = 6/C$
$0.255 = 6/C$
$C = 23.5 \approx 24$ cycles

APPENDIX G

Common Primers for Sequencing, Promoters, Housekeeping Genes, and Reporters

T3 promoter primer
AATTAACCCTCACTAAAGGG

T3 sequencing primer
ATTAACCCTCACTAAAGGGA

T7 promoter primer
TAATACGACTCACTATAGGG

T7 sequencing primer
TAATACGACTCACTATAGGG

T7 terminator primer
GCTAGTTATTGCTCAGCGG

C-terminal T7 tag
GGATCTACGTAATACGACTCACTATAG

SP6 promoter primer
TACGATTTAGGTGACACTATAG

SP6 sequencing primer
ATTTAGGTGACACTATAG

CMV forward (immediate early gene forward)
GGTCTATATAAGCAGAGCTGGT

ROSA26 promoter
F: AAAGTCGCTCTGAGTTGTTAT
R: GGAGCGGGAGAAATGGATATG

RSV long-terminal repeat forward
CGCCATTTGACCATTCA

SV40 primer for forward sequencing
TATTTATGCAGAGGCCGAGG

SV40 primer for reverse sequencing
GAAATTTGTGATGCTATTGC

Lambda gt10 primers
F: AGCAAGTTCAGCCTGGTTAAG
R: CTTATGAGTATTTCTTCCAGGGTA

Lambda gt11 primers
F: GGTGGCGACGACTCCTGGAGCCCG
R: TTGACACCAGACCAACTGGTAATG

M13 universal sequencing
F: GTTGTAAAACGACGGCCAGT
R: CAGGAAACAGCTATGACC

BGH reverse
TAGAAGGCACAGTCGAGG

LKO.1 5′ (U6 promoter forward)
GACTATCATATGCTTACCGT

pBABE 5′ for forward sequencing
CTTTATCCAGCCCTCAC

pGEX 5′ for forward sequencing
GGGCTGGCAAGCCACGTTTGGTG

pGEX 3′ for reverse sequencing
CCGGGAGCTGCATGTGTCAGAGG

EGFP-C plasmids' forward sequencing
CATGGTCCTGCTGGAGTTCGTG
3′-end of EGFP ORF

EGFP-N plasmids' reverse sequencing
CGTCGCCGTCCAGCTCGACCAG
5′-end of EGFP ORF

mCherry-F for forward sequencing
CCCCGTAATGCAGAAGAAGA
3′-end of mCherry

mCherry-R for reverse sequencing
TTGGTCACCTTCAGCTTGG
5′-end of mCherry

IRES-F for forward sequencing
TGGCTCTCCTCAAGCGTATT
3'-end of IRES

IRES-R for reverse sequencing
CCTCACATTGCCAAAAGACG
5'-end of IRES

MSCV
CCCTTGAACCTCCTCGTTCGACC

Firefly luciferase forward sequencing primer
GGATAGAATGGCGCCGG

Renilla luciferase forward sequencing primer
CCAGGATTCTTTTCCAATGC

Neomycin
F: CTTGGGTGGAGAGGCTATTC
R: AGGTGAGATGACAGGAGATC

RT-PCR primers for housekeeping genes:

Human GAPDH
F: TCTTCTTTTGCGTCGCCAG
R: AGCCCCAGCCTTCTCCA

Human beta-actin
F: CTGGGACGACATGGAGAAAA
R: AAGGAAGGCTGGAAGAGTGC

Human 18S RNA
F: CAGCCACCCGAGATTGAGCA
R: TAGTAGCGACGGGCGGTGTG

Mouse GAPDH
F: AGGCCGGTGCTGAGTATGTC
R: TGCCTGCTTCACCACCTTCT

Mouse beta-actin
F: TTCTTTGCAGCTCCTTCGTTGCCG
R: TGGATGGCTACGTACATGGCTGGG

Mouse 18S RNA
F: AGGGGAGAGCGGGTAAGAGA
R: GGACAGGACTAGGCGGAACA

APPENDIX H

Tags

Nucleotide and amino acid sequences for common tags and the monoclonal antibodies that recognize the tags

Tag	Tag nucleotide sequence (for N-terminal primer)	Tag amino acid sequence	mAb
HIS	CATCATCACCATCACCAT	HHHHHH	HIS-1
HA	TACCCATACGAC-GTCCCAGACTACGCT	YPYDVPDYA	12CA5
FLAG	GATTACAAGGATGACGAT-GACAAG	DYKDDDDK	M1, M2, and M5
V5	GGTAAGCCTATCCCTA-ACCCTCTCCTCGGTCTC-GATTCTACG	GKPIPNPLL-GLDST	V5-10
c-myc	GAACAAAAACTTATTTCT-GAAGAAGATCTG	EQKLISEEDL	9E10
VSV-G	TACACTGATATCGAAATGAAC-CGCCTGGGTAAG	YTDIEMN-RLGK	P5D4

mAb, Monoclonal antibody.

APPENDIX I

Thermostable DNA polymerases

Various thermostable DNA polymerases and their PCR-related properties

Name	Source organism	Half-life at 95°C (min)	5′–3′ exonuclease	3′–5′ exonuclease (proofreading)	DNA ends	Rate (nt/s)
Taq	*Thermus aquaticus*	40	+	−	3′ A overhang	75
Tfl	*Thermus flavus*	Unknown	−	−	Unknown	Unknown
Tli (Vent)	*Thermus litoris*	400	−	+	Blunt	67
Tma	*Thermus maritima*	50–60	−	+	Blunt	Unknown
Tth	*Thermus thermophilus*	20	+	−	3′ A overhang	60
Pfu	*Pyrococcus furiosus*	120–130	−	+	Blunt	60
Pwo	*Pyrococcus woesei*	Unknown	−	+	Blunt	Unknown

Example of a Daily PCR Log

Date: 24/06/2015

PCR master mix (100 μL):

dH$_2$O	→ 82
10× Buffer (Platinum, Invitrogen)	→ 10
50 mM MgCl$_2$	→ 2
dNTP (12.5 mM)	→ 2
actin-F (25 μM)	→ 1
actin-R (25 μM)	→ 1
Template	→ 1
Enzyme (Taq, Platinum)	→ 1

Program AN1:
94°C, 4 min,
94°C, 30 s
T = 70, G = 10; 30 s
72°C, 3 min
Repeat 35×
72°C, 10 min
4°C, 10 min
Results: No bands. Excess of template is evident.
Conclusions: Repeat with half the amount of the template.

Date: 25/06/2015

PCR master mix (100 μL):

dH$_2$O	→ 82
10× Buffer (Advantage)	→ 10
50 mM MgCl$_2$	→ 2
dNTP (12.5 mM)	→ 2
PM-F (25 μM)	→ 1
PM-R (25 μM)	→ 1
RT-PCR mix	→ 1
Enzyme (Advantage cDNA)	→ 0.5

PCR master mix (100 μL):

dH$_2$O	→ 82
10× Buffer (Advantage)	→ 10
50 mM MgCl$_2$	→ 2
dNTP (12.5 mM)	→ 2
CR1-F (25 μM)	→ 1
CR1-R (25 μM)	→ 1
pCEP4-RAF-1	→ 0.5
Enzyme (Advantage cDNA)	→ 0.5

Program AN1:
94°C, 4 min
94°C, 30 s
T = 65, G = 10; 30 s
72°C, 3 min
Repeat 30×
72°C, 10 min
4°C, 10 min
Results: Band sizes are correct for RAF-1 domains, but not for PM.

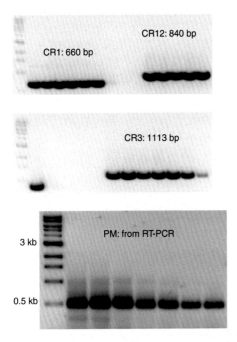

Conclusions: Cut out the domain bands and purify from the gel (Qiagen kit). Use for TOPO-cloning (pTrcHis). Redo the PM PCR with various [Mg^{++}].

APPENDIX K

Common Buffers and Solutions

50× TAE buffer (1 L)

242 g	Tris base
57 mL	Glacial acetic acid
10 mL	0.5 M EDTA, pH 8.0

EDTA, Ethylenediaminetetraacetic acid.

Add dH_2O to 1 L.

10× TBE buffer (1 L)

108 g	Tris base
55 g	Boric acid
40 mL	0.5 M EDTA, pH 8.0

Add dH_2O to 1 L.

5× agarose loading dye (10 mL)
Mix in this order:

3.6 mL	dH_2O
0.2 mL	0.5 M Tris–HCl, pH 7.6
1 mL	0.5 M EDTA
5 mL	100% glycerol
0.25 mL	1% BPB, 1% XC

BPB, bromophenol blue; XC, xylene cyanol.

10× TNE buffer (1 L)
Pour 800 mL of dH_2O into a 1 L beaker and add:

12.11 g	Tris base
3.72 g	EDTA
116.9 g	NaCl

Adjust pH to 7.4 with HCl.
Add dH_2O to1 L.
Filter through 0.2–0.45 µm filters.

APPENDIX L

Useful Numbers

1 μg of human genomic DNA = 3.04×10^5 molecules

1 μg of *Escherichia coli* genomic DNA = 2×10^8 molecules

1 μg of 1-kb dsDNA = 9.12×10^{11} molecules

1 μg of 1-kb RNA = 1.77×10^{12} molecules

APPENDIX M

Yeast Colony PCR

1. Pick a colony into 25 µL of 20 mM NaOH.
2. Vortex or pipette to resuspend.
3. Heat at 95°C for 5 min.
4. Use 1 µL for a 25-µL PCR reaction.

USEFUL WEBSITES

PCR resources:

- http://www.idtdna.com/SciTools/SciTools.aspx
- http://www.dnalc.org/shockwave/pcranwhole.html
- http://www.horizonpress.com/gateway/pcr-protocols.html
- http://www.sanger.ac.uk/Teams/Team53/psub/pcr/primer.shtml
- http://www.changbioscience.com/primo/pcr/index.htm

Online tools:

- BLAST: http://blast.ncbi.nlm.nih.gov/Blast.cgi
- Pairwise alignment: http://www.ncbi.nlm.nih.gov/blast/bl2seq/wblast2.cgi
- In silico translation: http://www.bioinformatics.org/sms2/translate.html
- Translation map: http://www.bioinformatics.org/sms2/trans_map.html
- Genetic code: http://www.geek.com/wp-content/uploads/2013/12/genetic-code.jpg
- Reverse complement: http://www.bioinformatics.org/sms2/rev_comp.html
- Primer design: http://www.invitrogen.com/content.cfm?pageid=9716
- Virtual restriction digestion: http://tools.neb.com/NEBcutter2/index.php
- Restriction enzyme information: http://www.neb.com
- Double digest finder: http://www.neb.com/tools-and-resources/interactive-tools/double-digest-finder

Others:

- www.ScienceLauncher.com/tools.html
- www.ProteinGuru.com

REFERENCES

[1] http://www.nobel.se.
[2] Mullis KB. The unusual origin of the polymerase chain reaction. Sci Am 1990;262:56–65.
[3] Giebel LB, Spritz RA. Site-directed mutagenesis using the double-stranded DNA fragment as a PCR primer. Nucleic Acids Res 1990;18:4947.
[4] Saike RK, Gelfand DH, Stoffel S, Scharf SJ, Higuchi R, Horu GT, et al. Primer-directed enzymatic amplification of DNA with a thermostable DNA polymerase. Science 1988;239:487–91.
[5] Flaman J-M, Frebourg T, Moreau V, Charbonnier R, Martin C, Ishioka C, et al. A rapid PCR fidelity assay. Nucleic Acids Res 1994;22:3259–60.
[6] Cline J, Braman JC, Hogrefe HH. PCR fidelity of Pfu DNA polymerase and other thermostable DNA polymerases. Nucleic Acids Res 1996;24:3546–51.
[7] Innis, M.A., Gelfand, D.H., Sninsky, J.J., White, T.J. (Eds.), 1990. PCR Protocols—A Guide to Methods and Applications. Academic Press.

FURTHER READING

[8] Eckert KA, Kunkel TA. High fidelity DNA synthesis by the *Thermus aquaticus* DNA polymerase. Nucleic Acids Res 1990;18:3739–44.
[9] Williams JF. Optimization strategies for the polymerase chain reaction. BioTechniques 1989;7:762–9.
[10] Ellsworth DL, Rittenhouse KD, Honeycutt RL. Artifactual variation in randomly amplified polymorphic DNA banding patterns. BioTechniques 1993;14:214–7.
[11] Gelfand DH, White TJ. Thermostable DNA polymerases. In: Innis MA, Gelfand DH, Sninsky JJ, White TJ, editors. PCR protocols. New York: Academic Press; 1990. p. 129–41.
[12] Innis MA, Gelfand DH. Optimization of PCRs. In: Innis MA, Gelfand DH, Sninsky JJ, White TJ, editors. PCR protocols. New York: Academic Press; 1990. p. 3–12.
[13] Sarkar G, Kapeiner S, Sommer SS. Formaqmide can drastically increase the specificity of PCR. Nucleic Acids Res 1990;18(24):7465.
[14] Smith KT, Long CM, Bowman B, Manos MM. Using cosolvents to enhance PCR amplification. Amplifications 1990;90(5):16–7.
[15] http://www.turnerbiosystems.com/doc/appnotes/s_0046.php.
[16] http://www.sigmaaldrich.com/content/dam/sigmaaldrich/docs/Sigma/General_Information/qpcr_technical_guide.pdf.
[17] http://www.idtdna.com/pages/docs/default-source/user-guides-and-protocols/primetime-qpcr-application-guide-3rd-ed-.pdf?sfvrsn=20.
[18] DNAPs: Abramson RD. Thermostable DNA polymerases. In: Innes MA, Gelfand DH, Sninsky JJ. editors. PCR strategies. San Diego, CA: Academic Press; 1995. pp. 39–57.
[19] Bartlett JMS, Stirling D, editors. Methods in molecular biology, vol. 226: PCR protocols. 2nd ed. Totowa, NJ: Humana Press Inc.
[20] Barnes WM. PCR amplification of up to 35-kb DNA with high fidelity and high yield from 1 bacteriophage templates. Proc Natl Acad Sci USA 1994;91:2216–20.
[21] Cheng S, Fockler C, Barnes WM, Higuchi R. Effective amplification of long targets from cloned inserts and human genomic DNA. Proc Natl Acad Sci USA 1994;91: 5695–9.
[22] Chou Q, Russell M, Birch D, Raymond J, Bloch W. Prevention of pre-PCR mispriming and primer dimerization improves low-copy-number amplifications. Nucleic Acids Res 1992;20:1717–23.

[23] D'aquila RT, Bechtel LJ, Videler JA, Eron JJ, Gorczyca P, Kaplan JC. Maximizing sensitivity and specificity of PCR by preamplification heating. Nucleic Acids Res 1991;19:3749.

[24] Don RH, Cox PT, Wainwright BJ, Baker K, Mattick JS. Touchdown PCR to circumvent spurious priming during gene amplification. Nucleic Acids Res 1991;19:4008.

[25] Frey B, Suppmann B. Demonstration of the Expand™ PCR system's greater fidelity and higher yields with a lacI-based PCR fidelity assay. Biochemica 1995;2:8–9.

[26] Kellogg DE, Rybalkin I, Chen S, Mukhamedova N, Vlasik T, Siebert P, et al. TaqStart antibody: Hotstart PCR facilitated by a neutralizing monoclonal antibody directed against Taq DNA polymerase. BioTechniques 1994;16:1134–7.

[27] Longo MC, Berninger MS, Hartley JL. Use of uracil DNA glycosylase to control carryover contamination in polymerase chain reactions. Gene 1990;93:3749.

[28] Nelson K, Brannan J, Kretz K. The fidelity of TaqPlusTM DNA polymerase in PCR. Strategies Mol Biol 1995;8:24–5.

[29] Kozak M. An analysis of 5'-noncoding sequences from 699 vertebrate messenger RNAs. Nucleic Acids Res. 1987;15:8125–48.

[30] Kozak M. An analysis of vertebrate mRNA sequences: intimations of translational control. J Cell Biol 1991;115:887–903.

[31] Kozak M. Downstream secondary structure facilitates recognition of initiator codons by eukaryotic ribosomes. Proc Natl Acad Sci USA 1990;87:8301–5.

[32] Roux KH. Optimization and troubleshooting in PCR. PCR Methods Appl 1995;4:5185–94.

[33] Sambrook J, Fritsch EF, Maniatis T. Molecular cloning: a laboratory manual. 2nd ed. Cold Spring Harbor, NY: Cold Spring Harbor Laboratory; 1989.

INDEX

Printed in the United States
By Bookmasters